The Art of Structures

The Art of Structures

Aurelio Muttoni

Introduction to the Functioning of Structures in Architecture

Translated from the Italian by Stephen Piccolo

E P F L Press

A Swiss academic publisher distributed by Routledge

www.routledge.com/builtenvironment

Taylor & Francis Group Ltd
2 Park Square, Milton Park
Abingdon, Oxford, OX14 4RN, UK

Routledge is an imprint of Taylor & Francis Group,
An informa business.

Simultaneously published in the USA and Canada by Routledge,
711 Third Avenue, New York, NY 10017

www.routledge.com

Library of Congress Cataloging-in-Publication Data
A catalog record for this book is available from the Library of Congress.

The publisher and author express their thanks to the Ecole Polytechnique Fédérale de Lausanne (EPFL) for its generous support towards the publication of this book.

Published under the editorial direction of Professor Manfred Hirt (EPFL).

EPFL Press

is an imprint owned by Presses polytechniques et universitaires romandes, a Swiss academic publishing company whose main purpose is to publish the teaching and research works of the Ecole Polytechnique Fédérale de Lausanne (EPFL) and other universities and institutions of higher learning.

Presses polytechniques et universitaires romandes
EPFL – Rolex Learning Center
Post office box 119
CH-1015 Lausanne, Switzerland
E-mail: ppur@epfl.ch
Phone: 021/693 21 30
Fax: 021/693 40 27

www.epflpress.org

Translated from the original *L'arte delle strutture*
© 2006 Accademia di architettura-Università della Svizzera italiana, Mendrisio

This book was translated by Stephen Piccolo
(www.transiting.en)

© 2011, First edition, EPFL Press, Lausanne (Switzerland)
ISBN 978-2-940222-38-4 (EPFL Press)
ISBN 978-0-415-61029-2 (Routledge)

Printed in Italy

Contents

Foreword   XI

Introduction   1

The path-way of structures 2
What is a load-bearing structure? 4
The purpose of a structure 4
Structure and architecture 5

Foreword and equilibrium, internal forces,
strength and stiffness   7

Loads that act on a structure 9
Gravitational forces and Newton's law of gravitation 9
Gravitational force on the surface of the earth 10
Force vectors, point of application and line of action 10
Free body diagram 11
Conditions of equilibrium of two forces 11
Forces that act on the surface of contact between two free
       bodies: action = reaction 12
Transmisson of a force and internal force 13
The compressive internal force and its quantification 13
Effect of compression on materials: compressive stress 14
The tensile internal force 15
Tensile stress 16
Effect of tension: elongation 16
Effect of compression:shortening 16
Linear behavior and elastic behavior 17
Stiffness 17
Stiffness of a structure subjected to tension or compression 18
Stiffness of the material 18
Elastic phase and plastic phase 19
Yield strength and strength 20
Mechanical behavior of steel 20
Modulus of elasticity E 20
Yield strenght f_y 21
Tensile strength f_t 21
Strain at failure ε_t 21
Tension and compression 22
Fragility and ductility 22
Concrete 23
Stone 23
Wood 24
Comparison of materials 24
Stiffness and strength 24
Dimensioning 26
Criterion of the serviceability limit state (SLS) 26
Criterion of the ultimate limit state (ULS) 27
Load factors 27
Factored loads and design value of the internal force 27
Resistance factors 27
Design strength 28
Fatigue 29
Equilibrium of more than two forces in a plane and in space 29
First condition of equilibrium 30
Polygon of forces or force polygon 30

Second condition of equilibrium 30
Point of application of a force and equilibrium 30
Angle of friction 31
Cremona diagram 31
Forces and internal forces 32

Cables   35

Structural diagrams 37
The span ℓ and the rise f 37
Supports 38
Direction of the internal force on the subsystem 38
Influence of the load 39
Influence of geometry 39
The ℓ/f ratio 40
Influence of the position of the load 40
Load in any direction 41
Cable with two vertical loads 41
Resultant cable 42
Cable with two non-vertical loads 42
Cable with multiple non-vertical loads 43
Parallel non-symmetrical loads 44
Auxiliary cable 44
Center of gravity 45
Funicular polygon 45
Distributed loads 45
Cable subjected to uniformly distributed loads 45
Catenary 47
Suspension bridges 47
Applications in architecture 48
Supports: pylons, anchors and other elements 49
Sizing of cables 49
Section of the cable in relation to the slenderness ratio ℓ/f 49
Quantity of material based on the slenderness ratio ℓ/f 50
Movements caused by variation of the intensity of loads 50
Deflections caused by permanent loads 51
Movements caused by variable loads 51
Deflections caused by temperature variation 52
Effect of horizontal movements of the supports
       on the geometry of the cable 52
Variation of the configuration of the loads 52
Movements caused by variable loads 53
Limiting displacements caused by variable loads 53
Increase of the permanent load 54
Solution with stiffening cable: cable beams 54
Solution with load-bearing cable and stabilizing cables 56
The cable with stiffening beam 57
Cables with flexural stiffness 58
Systems with combined cables 59
Cable-stayed systems 59

Cable networks, tents and membranes   61

Systems of cables in space 63
Cable networks 64
Tents and membranes 65
Pneumatic membranes 67
High-pressure pneumatic membranes 68

Arches 69

Structures under compression 71
Cases with multiple loads or distributed loads: arches 72
Parabolic arches 72
Catenary arches 72
Similarity between cables and arches 73
Influence of variable loads 73
Instability of arches 74
Provisions to stabilize arches 74
Addition of stabilizing bars 74
Insertion of a stiffening beam 75
Stiffening the arches by increasing the thickness 75
Line of action of the internal forces 76
Possible lines of action of internal forces inside an arch 78
Statically indeterminate and determinate arches 78
Three-hinged arches 79
Optimal form of a three-hinged arch 79
Required thickness of a three-hinged arch subjected only
    to compression 80
Arches constructed with materials resistant to tension 81
Arches whose form does not correspond to that of the
    funicular polygon of permanent loads 82
Two-hinged arches, the ideal form 82
Arches with one hinge 83
Arches without hinges 83
Semi-circular masonry arches 84

Vaults, domes and shells 87

Arches as roofing elements 89
Barrel vaults 89
Groin vaults 91
Fan vaults 92
Pavilion vaults 94
Domes 95
Effective functioning of domes 95
Domes with a central skylight opening 96
Steel domes 97
Crossed arches 97
Domes composed of arches and rings 98
Shape of domes and the stresses 99
Conical domes 100
Hyperboloids of revolution 101
Carrying of non-symmetrical horizontal or vertical loads 103
Geodesic domes 104
Grid domes 104
Shells and arbitrary domes 105
Downward double-curvature shells 106
Shells with upward or downward curvature, the hyperbolic
    paraboloid 107
Monkey saddle surfaces 108
Composed arbitrary shells 108
Cylindrical shells 109
Gridshell structures 110

Arch-cables 111

Carrying the horizontal thrust component 113
Arches with ties 114
Fixed and sliding supports 114
Design and analysis of arches with ties 115
Cable-strut compositions 116
Arch and cable compositions 117
Arch-cables 118
Stabilization of the arch and carrying of variable loads 118
Arch-cable cantilevers 119
Cable-stayed systems 120

Trusses 123

Solution to the problem of deformability and stability
    by the addition of supplementary bars 125
Trusses 125
Analysis of trusses 125
Unstable, statically determinate and statically indeterminate
    systems 128
Generation of trusses 129
General analysis of trusses 129
Upper chord, lower chord and diagonals 132
Influence of the height and the span on the internal forces
    in trusses 132
Complete analysis of a truss 133
Bending moments 134
Identification of the bars under largest stress in the chords 135
Targeted analysis of bars in the chords of trusses 135
Analysis of the diagonals and their functioning 137
Shear force 137
Identification of diagonals with the largest internal force 138
Identification of diagonals under tension and those under
    compression 138
Qualitative analysis of a truss 139
Possible configurations of the diagonals 139
V diagonals 139
N diagonals 140
X diagonals 140
K diagonals 142
Forms of trusses 145
Form and structural efficiency 146
Influence of form on structural stiffness 148
Cantilevers and towers with multiple loads 149
Towers 150
Trusses with cantilevers 151
Gerber beams 152
Trusses for other structural forms 153

Space trusses 155

Composition of trusses to support a roof 157
Lattice trusses 157
Space trusses 158
Vaults and domes composed of trusses 159

Beams 161

Internal forces in the middle zone of beams 163
Internal force in zones in tension and in compression 163
Reinforced concrete beams 164
Simple bending of a beam 165
Bending and curvature 165
Strength of beams subjected to bending 165
Influence of the dimensions of a rectangular beam on its
    strength 166
Influence of the dimensions of a rectangular beam on its
    stiffness 168
The most efficient sections: wide-flange sections 169
Influence of the dimensions of a wide-flange beam on its
    bending strength and stiffness 170
Behavior of a wide-flange beam with vertical flanges 171
The efficiency of a section 172
Form, section and structural efficiency 172
Simple beams with concentrated and distributed loads 174
Cantilevers 175
Beams with cantilevers 176
Gerber beams 178
Continuous beams 179
Bi-clamped beams 180
Zones of greater or lesser internal forces in beams 181

Frames 183

Two-hinged frames 185
Three-hinged frames 186
Form and stresses 187
Side-by-side frames 189
Stacked frames 191
Multistory side-by-side frames 192
Vierendeel beams 192

Deep beams and walls 195

Wall beams 197
Deep wall over several storeys 197
Deep walls in space 199
Folded plate structures 200

Ribbed slabs, beam grids and slabs 201

Composition of beams to support a flat area 203
Beam grids 204
Slabs 205
One-way slabs 205

Choice of slab thickness.. 206
Influence of the type of support on slab behavior 208
Continuous slabs 209
Functioning with concentrated loads 209
Two-way slabs 209
Equivalent spans for two-way slabs 210
Slabs on columns 211
Mushroom slabs 212
Flat slabs 212
Choice of thickness of slabs supported by columns 213

Stability of compressed members 215

Strength of a rod in compression 218
How to make a column stable 219
Influence of column height on compressive strength 220
Influence of the constraints, effective length 221
Influence of the stiffness of the material on the critical load
    of a column 224
Influence of the dimensions of the section................................. 225
Influence of the shape of the section 226
Choice of sections 227
Local buckling 227
Trusses and Vierendeel columns... 228
Variable-section columns 228

Appendices 229

Appendix 1. Analytical determination of the funicular
    curve with distributed load 231
Appendix 2. Analytical expression of the conditions
    of equilibrium 232
Appendix 3. Axial force, shear force and bending moment 234

Glossary 237

Bibliography 249

Photography credits 255

Index 259

i-structures 267

The concept of structure has always been a fundamental aspect of building. Until the Renaissance the statics of constructions was based solely on experience, intuition, experimentation with models and empirical rules, but the scientific revolution transformed this discipline into a true science. Since the mid-1700s it has been possible to calculate structures, analyzing their mechanical behavior. Their most efficient form can be determined by mathematical means, and the measures required to ensure their strength and stability can be set by comparing the internal forces with the strength of the materials. Through the technological developments and new materials that have emerged during the industrial revolution, the science of construction has made a range of new structural solutions possible. This phase required greater specialization, and the builder was replaced by two professional roles: the architect and the engineer.

For the engineer the situation that sprang from necessity and permitted extraordinary creative evolution has, over time, also revealed its limits. Structural analysis and calculation have become increasingly precise and detailed, and these improvements have led to more daring and more efficient structures, but all this has unfortunately had its price in terms of the conceptual design of structures, bringing about a gradual but inexorable weakening of the creative side of the work. Also for the architect, the separation of these disciplines has had advantages and drawbacks. The increasing difficulty of understanding structures has undoubtedly led to compromises.

In recent decades an attempt has been made to rectify this situation. Certainly, the solution is not to return to the way things were done in the past. The separation of the professions, based on real necessity, cannot be reversed. To resolve increasingly complex problems, the only path is that of dialogue and collaboration between different professional figures.

In order to collaborate and to design together it is indispensable to have shared interests, to use the same language and, above all, to understand each other. This book on structures in architecture is an attempt to make a contribution to this mutual understanding.

In concrete terms, it will help readers to understand the functioning of load-bearing structures; in practice, this means how loads are carried and transmitted to the ground.

Therefore the book focuses on comprehension. The study of the acting loads, the determination of forces and analysis of internal forces are also geared toward comprehension of how structures function. To facilitate matters, the approach will above all be intuitive. The fundamentals of equilibrium and structural functioning are explained on the basis of everyday experience. In this perspective the equilibrium of the human body, which we have experienced since we took our first steps, represents a valid example. The method applied here is effectively very different from the conventional approach based on the logical deduction of the laws of mechanics, statics and strength of materials. We will use the tools of graphic statics, keeping the use of analytical calculations to a minimum. A similar approach has already been described in a book written by Joseph Schwartz and Bruno Thürlimann in the mid 1980s on the design and dimensioning of structures in reinforced concrete.

This text was developed from a course specially conceived when the Architecture Academy of Mendrisio was founded in 1996, in an attempt to develop a true course on statics for architects, rather than a simplification of the classic approach used for training engineers. For an architect, understanding structural functioning is useful in the design of structures. Concretely, the idea is to learn founded how to choose an efficient structural type and the most suitable materials, to determine its statically correct form, to understand which zones are subjected to the greatest stress and to develop details in the best possible way. The study covers only the most important aspects of the calculation of internal forces and dimensioning of structural members, in order to facilitate dialogue with engineers.

I think these themes should also be of interest to engineers. Their training is still highly influenced by the approach developed in the first engineering schools, toward the end of the 1700s. As the Encyclopedists proposed, the science of construction can be seen as an application of mechanics, and this is but one chapter of physics. The result is a logical-deductive teaching whose aim is to provide the tools necessary for analyzing the stresses in structures and sizing their main parts. But unfortunately knowing how to calculate and dimension does not necessarily mean that one understands the functioning, or knows how to design a structure. So the approach taken by this book represents a necessary complement to classical teaching.

I would like to thank my assistants at the Architecture Academy: Stefano Guandalini, Paolo De Giorgi, Andrea Pedrazzinni and Patrizia Pasinelli who have accompanied me from the first time the course was given to the writing of this book.

Since 2003, this course has been taught to students in architecture and civil engineering of the Swiss Federal Institute of Technology in Lausanne (EPFL) and for some years now to the students in architecture of the Swiss Federal Institute of Technology in Zurich (ETH). It is in this context that I decided to have the book translated, first into French (2004) and now in English. I would like to thank both Olivier Burdet and Joseph Schwartz for their support and the constructive remarks as well Olivier Babel, Christophe Borlat and Frederick Fenter for their work at the *Presses polytechniques et universitaires romandes*, publisher of the books in French and English.

Aurelio Muttoni

Vous souhaitez être avertis de nos parutions ?

○ **par email**: connectez vous sur notre site *www.ppur.org*, et cliquez sur le lien «recevoir notre documentation».

○ **par courrier**: renvoyez-nous cette carte après l'avoir complétée au recto et au verso.
Valable pour: ○ Suisse et CEPT: particuliers et collectivités
             ○ Reste du monde: collectivités seulement

Société/Organisme/Université/Faculté/Ecole

Nom/Prénom

Fonction/Profession

Adresse complète, sans abréviation

Ville: Pays:

NPA:

PPUR
EPFL - Centre Midi
CH-1015 Lausanne

www.ppur.com

DEMANDE DE DOCUMENTATION

Je désire recevoir gratuitement:

- ○ Le catalogue du Traité des Matériaux ___ ex.
- ○ Le catalogue du Traité de Génie Civil ___ ex.

A parution, je désire recevoir gratuitement votre documentation dans les domaines suivants:
(Ne cocher les matières 01 ou 07 que si vous êtes intéressé(e) par toutes les sous-matières)

01 ○ Electricité, électronique, électrotechnique
   ○ Microélectronique
   ○ Electromécanique
   ○ Production et distribution d'énergie électrique
   ○ Traitement du signal
   ○ Télécommunications
   ○ Automatique, électronique de puissance
   ○ Théorie et modèles de l'électricité
02 ○ Génie civil
03 ○ Méthodes mathématiques pour l'ingénieur
04 ○ Gestion
05 ○ Sociologie du développement urbain et régional
06 ○ Droit économique
07 ○ Informatique
   ○ Matériel informatique
   ○ Langages et programmation
   ○ Génie logiciel
   ○ Informatique industrielle
   ○ Téléinformatique
   ○ Infographie
   ○ Intelligence artificielle
08 ○ Systèmes de communication
09 ○ Mécanique appliquée
10 ○ Architecture
12 ○ Collection Le savoir suisse

15 ○ Economie
16 ○ Direction d'entreprise
18 ○ Collection Focus Sciences
22 ○ Management of technology
23 ○ Histoire des sciences
29 ○ Mécanique des sols et roches
30 ○ Thermique
31 ○ Mathématiques
33 ○ Photochimie
34 ○ Optique
35 ○ Matériaux
36 ○ Chimie macromoléculaire
37 ○ Atlas de construction
38 ○ Génie de l'environnement
39 ○ Chimie appliquée
40 ○ Géologie
42 ○ Mécanique des fluides, hydraulique
43 ○ Géostatistique, cartographie
44 ○ Microtechnique
45 ○ Physique théorique
47 ○ Biologie
48 ○ Physique générale, expérimentale et appliquée
46 ○ Recherche opérationnelle
50 ○ Sciences forensiques
55 ○ Sciences de la vie

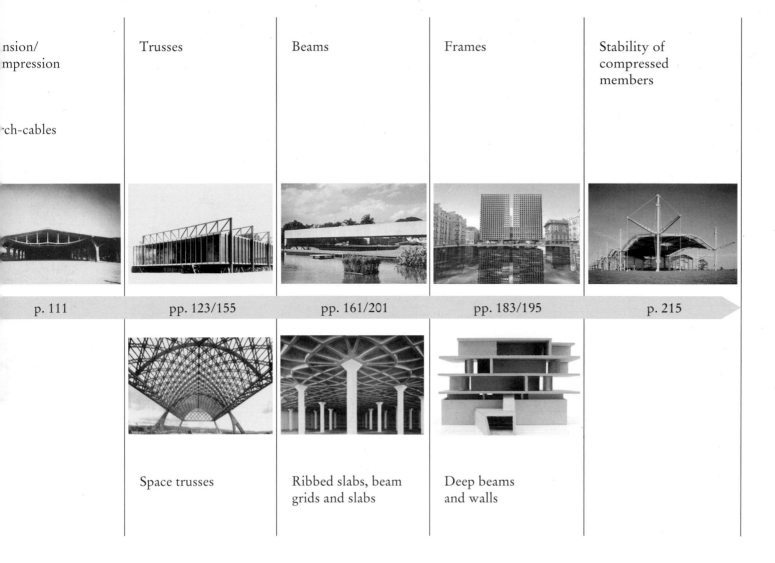

nsion/ mpression	Trusses	Beams	Frames	Stability of compressed members
rch-cables				

p. 111　　　pp. 123/155　　　pp. 161/201　　　pp. 183/195　　　p. 215

	Space trusses	Ribbed slabs, beam grids and slabs	Deep beams and walls	

he next step is to replace the bars of the trusses with extensive ones of material subjected to tension and compression. This enerates structures that function in a way very similar to usses, but with very different forms, because beams are volved. Composing beams in space we obtain grids, while by rther distributing the material we obtain slabs.

n particular, we will analyze the functioning of beams with a naller span with respect to height (deep beams and load-bear-g walls), discussing their similarities and differences with slen-er beams.

e will also take a look at frames, which can be interpreted as e composition of some elements already studied: columns, eams and arches.

Finally, we will examine beams subjected to a strong compression in their longitudinal direction (compressed elements and stability): these will be studied paying particular attention to the phenomenon of elastic instability due to the interaction between internal forces and deformation.

The path of structures also functions as a conceptual index. The use of the term "path" underscores the continuity of the presentation.
Note that the functioning of all structures, from the simplest to the most complex, can be described by indicating the zones subjected to tension (represented by continuous lines) and the zones subjected to compression (represented by dotted lines).

What is a load-bearing structure?

The term structure has different meanings. From our point of view, what counts is the assembly of elements that constitutes the skeleton or framework of a construction.

To be even more precise, we should talk about a *load-bearing structure*. This term, in building and other similar construction techniques, indicates the whole of the parts that have a load-carrying function.

On observing any building from the outside or the inside, it is usually quite easy to recognize at least part of the load-bearing structure. In the example shown here, the structure is clearly visible, and it is easy to distinguish a number of structural elements. For example, we can see vertical columns that have the task of transmitting loads to the ground, a series of horizontal trusses that support the floors of the building and transmit loads to the columns, bars arranged in X configurations and bars connecting the trusses to stabilize the construction and pick up the horizontal pressures caused by wind and earthquakes, also trusses attached to the columns and connected to other vertical bars, whose function we will learn later on, and other less obvious structural parts.

Upon closer observation, we can recognize vertical beams that make the facade more rigid and pick-up wind pressure, and a system of bars to which the escalator is attached, while the beams that constitute the structure of the escalator are part of the secondary structure.

At this point we may well wonder what is *not* part of the structure. An example is the air ducts we see on the roof, although they too have a structure capable of supporting their weight and standing up to wind pressure.

Georges Pompidou center in Paris, France, 1977, arch. R. Piano and R. Rogers, eng. P. Rice (bureau Ove Arup)

The purpose of a structure

The purpose of a structure is connected with its use and its architectural function. To simplify, we can indicate three possible main purposes of a structure:
— to enclose, cover or protect a space;
— to create a surface useful for other purposes (for example, a floor, a structure that supports a parking area, the bridge over which a road passes);
— to resist loads or to support something (a support wall that resists the pressure of the ground; a pylon that supports a power line; a chair, a table).

So the function of support and the capacity to resist loads is not necessarily the main purpose of a structure. All struc-

Exemple of a structure that cover un space: supreme court in Brasilia, Brazil, 1958, arch. O. Niemeyer, eng. J. Cardoso

Exemple of a structure that creates a useful surface; walkway in Durham, UK, 1965, eng. O. Arup & Partners

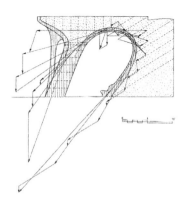

Exemple of structure that supports the ground: Güell Park in Barcelona, Spain, arch. A. Gaudi, 1900-1904

Load-bearing structure as the main element of the facade, Centre National de Chorégraphie, Aix-en-Provence, France, 1999-2006, arch. Rudy Ricciotti, eng. Serge Voline

tures inevitably have mass. As a result, the capacity of a structure to "carry" its own weight is a constant, qualifying characteristic.

In addition to the purposes we have presented here, load-bearing structures often serve other functions, and it is for this reason that they become an important element in architecture. Indeed, the load-bearing element can organize or "structure" space through the frame it imposes. In other cases, the presence of the structure can be exaggerated and even become a fundamental element of the space.

Structure and architecture

Forces and equilibrium, internal forces, strength and stiffness

Centre Georges Pompidou in Paris, France, 1977, arch. R. Piano and R. Rogers, eng. P. Rice (Ove Arup Engineer)

Loads that act on a structure

The external forces that act on a structure are defined as *loads*. If we consider the example shown here already seen on page 4, we can immediately see that loads can essentially be subdivided into four groups:

1. permanent loads (the weight of the structures themselves and the non-structural elements that remain constant over time);
2. variable loads (the weight of persons, furniture, snow, etc.);
3. wind pressure;
4. inertial forces caused by accelerations of mass (earthquakes, impacts, etc.).

Gravitational forces and Newton's law of gravitation

For the moment we will focus on the forces of the first two groups, which in physical terms are *gravitational forces*. A person, a chair, a table or a part of the structure (any mass) is subject to the attraction of the earth. In other words, a mass placed on the surface of the earth is subject, first of all, to the gravitational force exercised by the earth itself.

We also know that the earth is subjected, in turn, to the gravitational attraction exercised by the masses found on its surface, for example by people. The earth and the person can be considered as two masses that attract each other reciprocally. According to *Newton's law of gravity*, the force the earth exercises on the person has the same intensity as the force the person exercises on the earth: it is proportional to the product of the two masses and inversely proportional to the square of the distance.

This relationship can be expressed by the following equation:

$$F_{2,1} = F_{1,2} = G \cdot \frac{m_1 \cdot m_2}{r^2}$$

$F_{2,1}$ is the force m_2 exercises on m_1.
$F_{1,2}$ is the force m_1 exercises on m_2.
r is the distance between the two masses.
G is the universal gravitational constant
$G = 6.67 \cdot 10^{-11}$ N $\cdot$ m^2/kg^2

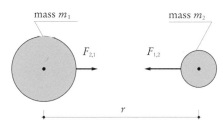

Gravitational forces between two masses m_1 and m_2 at a distance r

Like all forces, *gravitational force* can be quantified in Newtons. A Newton is equal to the amount of force required to give a mass of one kilogram an acceleration of one meter per second squared:
1 Newton = 1 N = 1 kg $\cdot$ m/s^2

Gravitational force on the surface of the earth

For all bodies placed on the surface of the earth, with $m_2 = 5{,}985 \cdot 10^{24}$ kg and $r = 6\,378\,000$ m, the following relationship applies:

$$F_{2,1} = F_{1,2} = m \cdot 9.81 \text{ with the units N} = \text{kg} \cdot \text{m/s}^2.$$

So 1 kg of mass has a weight of 9.81 N, approximately 10 N, while a person with a mass of 70 kg has a weight of 687 N, about 700 N, equivalent to 0.7 kN.

Note that even after climbing to the top of Mt. Everest, the weight of the same person will remain very similar, because the distance from the center of the earth has not been significantly increased: with $r \cong 6\,378\,000 + 8\,800$ m the gravitational force is

$$F_{2,1} = 6.67 \cdot 10^{-11} \cdot 70 \cdot 5.958 \cdot 10^{24}/6\,386\,800^2 = 685 \text{ N}.$$

If we compare this force to the 687 N we obtained by calculating the force acting on the same person at sea level, we see that the difference is minimal. Other differences of the same almost negligible order are also caused by movement on the earth's surface at a constant altitude. This effect is caused by the fact that the earth is not a perfect sphere and does not have a uniform distribution of density.

The same person, however, will have a very different weight if he visits another planet. For example, on the moon a person of 70 kg is subjected to a gravitational force of

$$F_{2,1} = 6.67 \cdot 10^{-11} \cdot 70 \cdot 7.35 \cdot 10^{22}/1\,738\,800^2 = 114 \text{ N}$$

equal to about one sixth of the force the same person undergoes on earth. For this reason, an astronaut who tries to walk on the surface of the moon cannot help but jump.

Force vectors, point of application and line of action

The forces we have examined all have *a point of application*. It coincides with the center of gravity (also known as the center of mass). As we proceed we will also learn how to determine the center of gravity of an arbitrary body.

Force can be represented as a vector defined by its point of application, its direction, its sense and its intensity. When representing multiple forces, we will draw the length of their vectors in proportion to the intensity of the force.

The point of application and the direction determine the *line of action of a force* (or *line of force*). In the case of gravita-

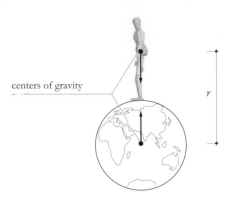

centers of gravity

r

Gravitational forces at work between the earth and a person on its surface

On the moon, a person has a very different weight

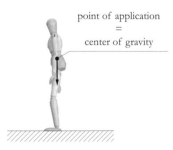

point of application
=
center of gravity

Center of gravity and vector representing the force-weight applied by the earth on a person

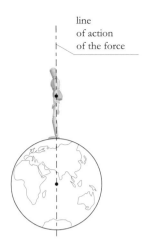

line
of action
of the force

Line of action of the force-weight

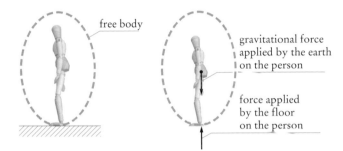

free body

gravitational force
applied by the earth
on the person

force applied
by the floor
on the person

"Person free body" and acting forces

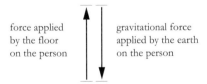

force applied
by the floor
on the person

gravitational force
applied by the earth
on the person

Comparison of the forces acting on the free body

tional forces, the line of action connects the two centers of gravity (that of the person and that of the earth).

Nevertheless, a person positioned on the surface of the earth or on a floor does not undergo only the gravitational force of the earth that draws the person downward. Our experience shows us that the floor plays a fundamental role, and we are well aware of what can happen if the floor is suddenly not there under our feet! The floor has an "active" function to support a person, thus exercising an upward force.

In order to represent this force, we have to isolate the person from the rest of the earth-person system. From now on we will give the name *"free body diagram"* to precisely this part of the system that we have isolated, together with all the forces at work there.

Free body diagram

We can easily imagine the conditions in which the two forces that act on the person will be in equilibrium. The forces must:

1. be of equal intensity;
2. act in the same direction, but in an opposite sense;
3. be placed along the same line of action.

Conditions of equilibrium of two forces

The first two conditions can be summed up in an equivalent condition: the two forces must vectorially cancel each other out.

In fact, the person can stay motionless on the floor only if the forces counter each other, producing a condition of equilibrium.

To understand the meaning of the condition that dictates that the two forces must act along the same line of action, we can try inclining our body so that the line of action of the gravitational force, which passes through its center of gravity, is shifted forward, reaching the tip of our toes. At this point the equilibrium will be precarious.

If we lean further forward, the line of action of the gravitational force no longer passes through our feet, and we fall down. This is because the force the floor exercises on us must act on the surface of contact between the feet and the floor; if the two forces can no longer have the same line of action, equilibrium will not be possible and we will fall forward.

Actually, even in this case equilibrium of the forces is still possible, but only if we introduce an inertial force capable of considering the acceleration of our body as it falls. In any case this approach, which concerns the dynamics of bodies, will not be explored in this book.

To analyze the forces that act on a person, we have isolated the free body represented by the person from the overall system (person + earth). Clearly, we can repeat the same analysis of the equilibrium with regard to the "earth" free body as well. In this case we still have two forces: the gravitational force the person exercises on the earth (upward), and a force the person exercises on the floor, i.e. on the earth (downward). To meet the conditions of equilibrium, the two forces, must again have the same intensity, as in the free body analyzed previously.

Since the gravitational force the person exercises on the earth ("earth free body") has the same intensity as the earth's attraction on the person ("man free body"), all four forces that act on the earth and on the person are of equal intensity.

Forces that act on the surface of contact between two free bodies: action = reaction

With the above example we have demonstrated that the forces that act on the surface of contact between two free bodies (the action of the floor on the person and the action of the person on the floor) must have the same intensity. To describe this situation, which always applies when we isolate two free bodies, we say that "the action corresponds to the reaction" (*actio = reactio*, according Newton's formula, who was the first to express this equation).

This rule implies the definition of the subject who exercises the action and undergoes the reaction, but it clearly also applies if we swap the subject with the object. To avoid possible confusion, it is always useful to define *who* exercises the force and *on* what (the force of the person on the floor / the force of the floor on the person).

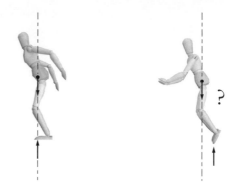

Situation of precarious equilibrium

Situation without static equilibrium

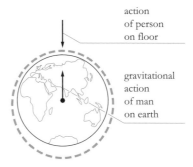

action of person on floor

gravitational action of man on earth

"Earth free body" and acting forces

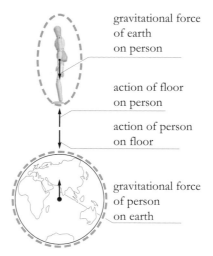

gravitational force of earth on person

action of floor on person

action of person on floor

gravitational force of person on earth

Combination of the two free bodies with their respective forces

Glo-Ball lamps, 1998, Jasper Morrison, table lamp, suspended lamp and lamp stand.

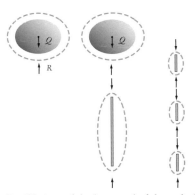

Equilibrium of the lamp and of the rod of the lamp stand

From these considerations, we are now ready to investigate our first structures. The lamp shown here, which can be placed on a table, on a rod or suspended from the ceiling will help us in this process. Carrying it or suspending it requires an additional element: a rod or a cable.

Let's first consider the case of the table lamp and the lamp stand. With the table lamp, we can proceed with the same analysis that we made for the person standing on the earth. A free body that includes only the lamp and isolates it from the table allows us to consider the contact force applied by the table on the lamp: the force R must be identical to the weight Q of the lamp. ($R = Q$).

We can of course do the same for the lamp stand. The force applied by the rod on the lamp must also be equal to the weight Q and from the standpoint of equilibrium, the fact that the lamp is screwed to the rod and that the free body diagram must cut the rod does not change anything. The only difference is that the upward force that we must introduce is not a contact force any more (between the table and the lamp), but represents the *internal force* in the material of the lamp's rod.

If we now assume that the weight of the rod is negligible with respect to that of the lamp, we can cut a second free body that only contains the rod. The same approach as previous allows one to demonstrate that the same forces act on the rod: a downward forces Q acting on the top of the rod element, and the same upward force acting on its lower end. Acting on the rod, therefore, are the downward force from the lamp and the upward force from the floor (or the base of the lamp stand, to be more accurate). We can thus interpret this configuration as the transmission of the lamp's force to the floor and vice-versa.

In other words, we can say that the rod is subjected to an internal force by the lamp and the floor, or that it is compressed between the lamp and the floor. We can of course cut several free bodies in the rod, and we will see that the forces acting on them are all identical. This means that the *internal force* is identical along the entire height of the rod, independently of the length of the element that we have isolated and of its cross section (should it be variable).

Transmisson of a force and internal force

The compressive internal force and its quantification

We can quantitatively define the *compressive internal force* acting within the rod by simply relating it to the force that the rod carries from one end to the other (in our case the weight of the lamp):

$N = -Q$ measured in N or in kN

where N is the internal force and Q the load acting on the rod. The minus sign in the equation derives from the sign convention, as it is usual to define compression as a negative quantity.

Effect of compression on materials: compressive stress

We have seen that the internal force carried by the rod is independent of its section. Of course this is true only when the weight of the rod is neglected. But we can easily guess that the cross-section of the rod will influence the level of stress on the material. To explain this concept, let us look at the following example.

In the two rods shown here, the one on the right has a section that is twice as large as the one on the left. As we have seen, the internal force is independent of the length of the rod and its section. The internal force in the two rods is therefore identical. Nevertheless, it is evident that if we divide the two rods into portions of material of the same size (for example, by vertically slicing the rod on the right), the two portions of the rod will be subjected to a force that has been halved with respect to that of the rod on the left (in this case we have assumed that the line of action of the internal force corresponds to the barycentric line of the rod). To take into account the fact that the effective stress on the material, unlike the force on the rod, depends not only on the compressive internal force but also on the cross-sectional area of the section, we can calculate the internal force per unit of area:

$\sigma = N/A$ measured in N/mm²

where N is the internal force and A is the area of the cross-section.

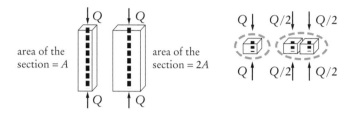

area of the section = A

area of the section = $2A$

Two rods with different sections subjected to the same force, free bodies of rod segments

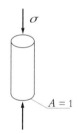

Compressive stress

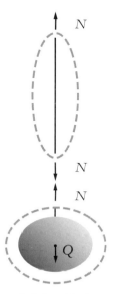

Equilibrium of the lamp and of the cable that supports it

This internal force per unit area is called stress (in this specific case, *compressive stress*), and it defines the amount of force on the material. Because the internal force N has a negative sign (–), σ will be negative. We need not attribute too much importance to this negative sign: it represents a mere convention that is useful to distinguish compressive stress from other types of stresses.

Let us now consider the case of the lamp suspended from the ceiling, assuming again that the weight of the cable is negligible with respect to that of the lamp.

The tensile internal force

We see first that on the free body that includes the lamp, two forces are again at play: the weight of the lamp and the internal force exerted by the cable on the lamp to support it. Since this internal force must again act upward to balance the weight (the downward attraction of the earth), it is pulling the lamp upwards instead of pushing it as in the previous case. The free body that includes the cable shows also that the internal forces are not the same as previously: the internal force acting on the cable at its lower end must act downwards and thus pull on the cable. This means that we have a *tensile internal force*. Indeed, we always have tensile internal forces when the vectors that represent them are pointing away from the section (cut of the material at the edge of the free body). Conversely, we have a compressive internal force when the vectors are pointing towards the section, as in the case of the rod of the lamp stand.

As in the column subjected to compression, again in this case we can quantify the *tensile internal force* by simply indicating the force that is transmitted from one end to the other:

$N = Q$ measured in Newtons or in kN

where N is the internal force and Q the load supported by the cable. By convention, a tensile internal force is indicated as positive. The + or – sign serves to distinguish the positive tensile internal force from the negative compressive internal force.

Tensile stress

We should note that the internal force N described in the previous chapter indicates the amount of force on the cable, but this alone does not supply any information about the effective stress on the material. To obtain this information we have to divide the internal force by the effective area of the section of the cable:

$\sigma = N/A$ measured in N/mm^2,

where N is the internal force measured in Newtons and A the area of the section measured in mm^2.
For this value, defined as *tensile stress*, we have the same expression we introduced to indicate the stress on the compressed material of the rod. The stresses σ are distinguished by a different sign: positive for tension and negative for compression.

Effect of tension: elongation

Let us now consider the effect of the internal force of tension on the cable. Experience teaches us that any tensile force has the effect of lengthening the object it acts on.

Effect of compression: shortening

Similarly, an element subjected to compression tends to become shorter.
If we are not satisfied with these trivial statements, we may wonder *how much* the cable is lengthened and *how much* the rod is shortened. Moreover, we can try to understand what factors influence these specific deformations (of lengthening and shortening):
  – clearly the internal force plays a fundamental role: the greater its intensity, the greater the deformation;
  – the type of material is also undoubtedly important: let us imagine a steel cable and an elastic material (with the same area and equal internal force);

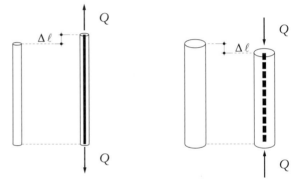

Elongation and shortening

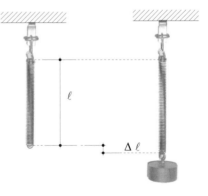

Steel spring subjected to a force of weight Q

- the dimensions (area of the cross-section and length) will also certainly have an influence on the deformations.

To quantify these influences we can perform an experiment. We can apply tension to a steel spring, for example by hanging a weight Q from it, and measure the elongation $\Delta\ell$.

Linear behavior and elastic behavior

Increasing the internal force N, which in our case corresponds to the load Q, also increases the elongation $\Delta\ell$. More precisely, if we double the internal force, we obtain twice as much elongation. This means that if we put these values into a diagram, we still obtain a straight line. In other words, we are looking at a *linear behavior*. As we will see, this does not necessarily apply to all materials, or at least not to any internal force applied to any material.

If we remove the weight, therefore removing the internal force, we return to the starting point with $\Delta\ell = 0$. So in this case we are dealing with perfectly reversible behavior. In such cases, we use the term *elastic behavior*.

Stiffness

If the behavior is linear and reversible, just one parameter will suffice to quantify the internal force-elongation relationship. This value, which represents the slope of the straight line in the diagram $N(\Delta\ell)$, is defined as the *stiffness* and corresponds to the relation $N/\Delta\ell$.

Now let us double the length of the spring by adding another spring with the same characteristics. With the help of the appropriate free bodies, we can immediately demonstrate that the two springs are subjected to the same internal force as in the case of the single spring we have just analyzed. In this case we will have to sum the elongations of each spring, so we will have a doubled elongation and a halved stiffness with respect to the case of the single spring. To generalize, we can say that the *stiffness is inversely proportional to the length ℓ*.

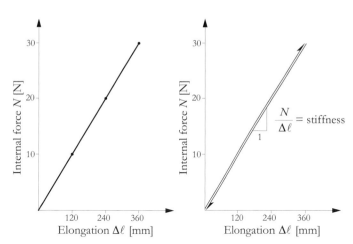

Internal force - elongation diagram, linear behavior

Elastic behavior and stiffness

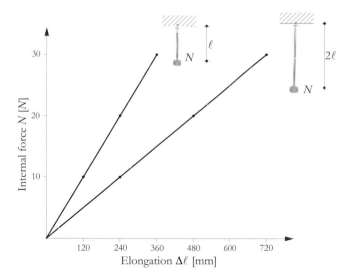

Influence of the length

To simulate the influence of the area of the section we can place two springs parallel to each other. In this case it is as if we had doubled the area A of the section of our element. Each of the two springs will clearly be pulled with half of the total internal force N, so the elongation $\Delta\ell$ will be halved with respect to the case of the single spring. The stiffness of the system, then, is doubled, so we can state that *the stiffness is directly proportional to the area of the section A.*

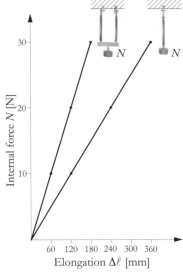

Influence of the area of the section

Stiffness of a structure subjected to tension or compression

To summarize: the stiffness of a simple structure subjected to tension can be defined by the relationship between the internal force N and the elongation $\Delta\ell$. This also depends on the material, and is directly proportional to the area A of the section and inversely proportional to the length $\Delta\ell$. Introducing a constant that defines the stiffness of the material (E, known as the *modulus of elasticity, elastic modulus* or *young's modulus*), we can write the following law:

$$\frac{N}{\Delta\ell} = \frac{E \cdot A}{\ell}$$

The expression is also valid for a structure subjected to compression. In that case N and $\Delta\ell$ (the shortening) will be negative.

Stiffness of the material

The equation we have just found can be transformed by putting $\Delta\ell$ as the numerator and A as the denominator to obtain

$$\frac{N}{A} = E \cdot \frac{\Delta\ell}{\ell}$$

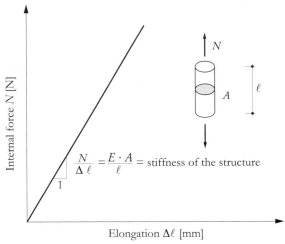

$$\frac{N}{\Delta\ell} = \frac{E \cdot A}{\ell} = \text{stiffness of the structure}$$

Mechanical behavior of the structure

As we have already seen, the first term, called stress σ, corresponds to the specific internal force on the material.
If we define $\Delta\ell/\ell$ as strain ε (deformation per unit of length), we will then have the following law:

$$\sigma = E \cdot \varepsilon \quad \text{or} \quad \sigma/\varepsilon = E$$

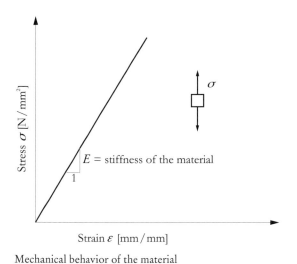

Stress σ [N/mm^2]

E = stiffness of the material

1

σ

Strain ε [mm/mm]

Mechanical behavior of the material

The second expression defines the stiffness of the material (*the modulus of elasticity*) as a relationship between the specific stress σ and the strain ε.

This expression refers to the material and is very similar to the one that defines the stiffness of the structure. Their graphic representations is also similar.

Note that in the case of the diagram referring to the structure, if we vary the area of the section and the length we naturally obtain different results. These parameters, on the other hand, are not present in the diagram that refers to the material.

Elastic phase and plastic phase

Experience shows that when we apply forces on materials beyond a certain limit, their behavior stops being elastic (i.e. reversible).

If we bend an iron wire without reaching this limit, by removing the force we can make the wire return to its initial form. If we exceed the limit, though, we will have an irreversible deformation. If we remove the force again, the wire will return toward its initial form but will not fully reach it. Later we will examine how, inside a bent wire, tensile internal forces are accompanied by compressive internal forces (bending phenomenon). For the moment, it is enough to know that identical behavior is produced in an element subjected to tension as in one subjected to compression.

Iron wire bent in the elastic phase

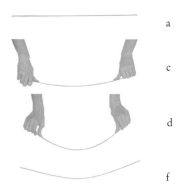

a

c

d

f

Iron wire bent beyond the elastic phase, irreversible deformation, plastic phase

If we represent relationship between the stress σ and the strain for iron or steel ε in a diagram, we obtain a complex relationship with multiple phases, which can all be described by straight lines. In the phase c-d-e we have irreversible strain and constant stress (horizontal segment in the diagram). This phase is called the *plastic phase*.

The phase e-f, in which the stress is reduced, is also characterized by a linear relationship, in which the slope is identical to that of the first elastic phase a-b-c. So we have the same modulus of elasticity E.

Although it has undergone a plastic deformation, if we load the material again we will follow the same line f-e until reaching the elastic limit beyond which there will be further (irreversible) plastic deformations.

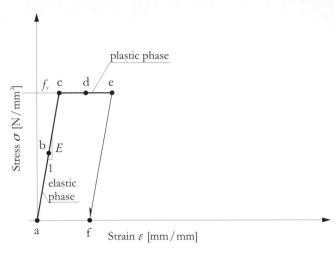

Stress-deformation diagram, elastic phase and plastic phase

Yield strength and strength

For iron, steel and all materials with similar characteristics, this limit is called *yield strength* f_y. If we extend the material further, we will reach a deformation beyond which the stress will begin to increase again. This phase is called *strain hardening*.

At a certain point we will reach the *strength* of the material f_t. A crack will appear that crosses all the material, leading to failure by separation.

Mechanical behavior of steel

The behavior of steel, which in the stress-strain diagram is characterized by two linear segments (elastic phase and plastic phase) and a curve (strain hardening phase), can be quantified with four parameters: modulus of elasticity E, yield strength f_y, tensile strength f_t, and deformation at failure ε_t.

Modulus of elasticity E

The modulus of elasticity E is constant for all types of steel: $E = 205000$ N/mm².

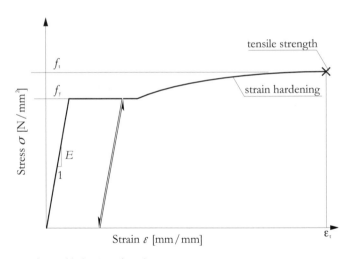

Mechanical behavior of steel

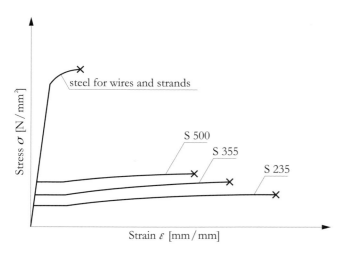

Stress-deformation diagrams for four types of steel under tension

Type of steel	E N/mm²	f_y N/mm²	f_t	ε_t
S 235	205000	235 *	360 *	0,210 *
S 355	205000	355 *	510 *	0,170 *
S 500	205000	500 *	580 *	0,140 *
steel for wires	160000÷	1410÷	1570÷	0,050 *
and strands	195000**	1670 *	1860 *	

* minimum values
** values reduced to take into account the effect of twisted wires

Mechanical values of the four most common types of steel

The yield strength f_y depends on the type of steel. It varies from a minimum of 235 N/mm² for the steel commonly used in steel constructions (which we will call S235), to a maximum of about 3000 N/mm² for the material of the strings of a pianoforte (spring steel). Note that spring steel does not have a true plastic phase with constant yield strength. For such cases it is customary to indicate the yield strength f_y as the stress that corresponds to a plastic deformation of 0.002 mm/mm (2 mm/m). Between these two extremes, we find high-strength steel, used in steel constructions (f_y = 355 N/mm², S355), reinforcing steel for use in reinforced concrete structures (f_y = 500 N/mm², S500), and spring steel, used for the wires and strands that form the cables of cable-stayed structures and the prestressing tendons used in pre-stressed reinforced concrete (f_y = 1410 – 1670 N/mm²).

Yield strength f_y

The tensile strength f_t is always greater than the yield strength and corresponds to the maximum stress reached before failure.

Tensile strength f_t

The strain at failure ε_t can reach very high levels. Since such large strains can be problematic for a structure, in dimensioning we usually consider only the plastic phase with the yield strength f_y into account, neglecting the increased stress that happens with strain hardening. For this reason, the type of steel to be used is indicated by specifying the yield strength. It is useful to take a look at the magnitude of strain at failure. For S235 steel commonly used in metal constructions, the failure strength is reached for an strain that corresponds to one fifth of the original length.

Strain at failure ε_t

Tension and compression

The values of the modulus of elasticity E and the yield stress f_y are identical for steel in tension and in compression. However the ways in which the tensile and compressive internal forces are transmitted inside the material are quite different. A piece of steel or any other material is capable of resisting compression even if it is subdivided into elements that transmit the internal force by mere contact. The dotted line we have chosen to represent compression metaphorically evokes this situation, with the individual rectangles indicating the blocks of the compressed material.

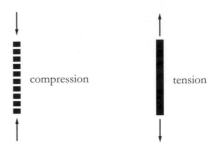

compression   tension

Graphic representation of compression and tension

In the case of a tensile force acting on the material, the material must be continuous or at least connected by another material that is resistant to tension (glue, or welding or porter). This is why from now on we will use a continuous line to represent tensile internal forces.

The example of a chain that transmits a tensile force might, at first glance, seem to contradict what has just been stated. Actually the links of the chain, though mainly under tension, contain a complicated set of zones under compression and tension, where the internal force is transmitted to contiguous links by contact under compression.

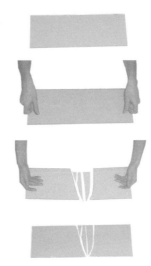

Fragility and ductility

Now let us try to deform a small pane of glass, just as we did before with the steel wire. Under small strains the glass will deform with linear-elastic behavior. But we know very well what will happen if we increase the deformation to the failure point: the glass will snap. Again in this case, then, we have a failure strength, which determines the *strength of the material* f_t.

The difference in behavior with respect to steel is clear: in the case of glass, the failure happens without the material having previously undergone (irreversible) plastic strain. This behavior is defined as *fragile*.

The other most common construction materials behave in ways that range between these two extremes:
- the ductile behavior of steel;
- the very fragile behavior of glass.

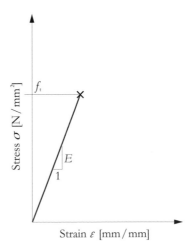

A pane of glass deformed slightly and then deformed to the failure point

Stress σ [N/mm²]

f_t

E
1

Strain ε [mm/mm]

Mechanical behavior of glass under tensile or compressive forces

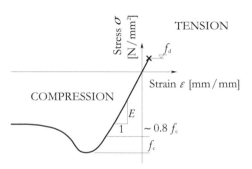

Mechanical behavior of concrete

Type of concrete	Modulus of elasticity E N/mm²	Tensile strength f_t N/mm²	Compressive strength f_c N/mm²	
Low-strength concrete	28 000÷34 000	1.1÷2.0	15 */ 10 **	concrete obtained with little cement and a lot of water
Concrete commonly used in construction	30 000÷36 000	1.5÷2.9	25 */ 17 **	normal concrete
Concrete for bridges	34 000÷42 000	2.2÷4.2	35 */ 23 **	
High-strength concrete	37 000÷44 000	2.9÷5.3	60 */ 40 **	concrete obtained with a lot of cement and less water

* minimum values
** design values

Mechanical values of four types of concrete

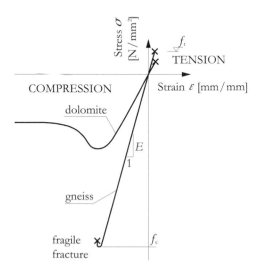

Mechanical behavior of stones

Concrete

If we analyze concrete, we see that its behavior depends essentially on the type of internal force applied. If subjected by tension, concrete behaves with great fragility, similar to that of glass. Failure happens by separation, due to a crack that appears perpendicular to the direction of the internal force.

Under compression, concrete behaves in a very different way. Beyond a certain limit, cracks appear parallel to the direction of the internal force, causing irreversible deformation. Yet the internal force can be increased to the point of reaching the compressive strength f_c, beyond which the spread of other cracks causes a reduction of strength.

By varying the content of the cement, the type of aggregate and, above all, the water content in the mix, concretes of different strengths can be obtained.

Stone

From a certain viewpoint, concrete can be thought of as a rock (conglomerate). In fact many stones have mechanical behavior very similar to that of concrete. There is also a wide range of strengths for different types of stone. In general, sedimentary stones (sandstone, conglomerate, dolomite) have strengths and a modulus of elasticity that are lower than those of metamorphic and crystalline stones (marble, gneiss, granite, basalt, etc.).

For both concrete and stone, the behavior of material under compression varies according to strength. Low-strength stones and concrete show rather large irreversible strains before and, above all, after reaching the point of compressive strength. For high-strength materials the behavior can, instead, become very fragile. Upon reaching compressive strength f_c, numerous cracks rapidly appear that cause explosive rupture, with the formation of shards.

Failure caused by tension, on the other hand, is always very fragile and happens through the formation of a single crack perpendicular to the internal force.

Wood

Wood, a biological material essentially made of cellulose, looks like a true structure when observed under a microscope. It is in fact composed of tubular structures with a slender wall, very similar to that of corrugated cardboard. The mechanical behavior of the material when loaded in the direction of growth is quite similar to that of other structural materials, with a significant linear elastic phase.

If subjected to tension beyond a certain limit, the cellulose structures tear, forming cracks, and the material loses its strength. If subjected to compression, the crushing and the instability of the thin walls of the small tubes cause irreversible deformations. Even after reaching the compressive strength f_c, wood can still be crushed with significant plastic deformation and a small loss of strength.

Different types of wood have different mechanical characteristics that depend, essentially, on their differing density (quantity of cellulose and quantity of openings with respect to the total volume).

Comparison of materials

Structural materials, then, have a wide range of different mechanical characteristics. The comparison shown here between concrete and steel subjected to tension is a good illustration of this situation.

Stiffness and strength

All materials, in any case, have an elasticity modulus E and a strength f_t (or f_c). Naturally these two characteristics can also be found in structures. As we have seen, the stiffness EA/ℓ of the structure corresponds to the modulus of elasticity E of the material.

Type of rock	Modulus of elasticity E N/mm²	Tensile strength f_t N/mm²	Compressive strength f_c N/mm²
Sandstone	6000÷20 000	1÷2	10÷60
Carrara marble	60 000÷90 000	2÷15	80÷130
Gneiss and granite	20 000÷50 000	2÷15	80÷180

Indicative values for certain types of stone

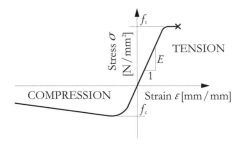

Mechanical behavior of wood under longitudinal force

Type of wood	Density kg/m³	Modulus of elasticity E N/mm²	Resistance to tension f_t N/mm²	Resistance to compression f_c N/mm²
Balsa-wood	80÷200	2500-6000	16-22	8-18
Fir:				
– without defects	400÷500	14 000	100	30
– construction wood		12 000	40	20
Beech	600÷750	15 000	130	50
Oak	600÷800	16 000	140	50

Stiffness and strength of wood, indicative values for stress parallel to direction of growth

	S500 steel	Ordinary concrete	Steel/concrete ratio
Modulus of elasticity E	210 000	~ 33 000	~ 6:1
Tensile strength f_t	500*	~ 2.5	~ 200:1
Compressive strength f_c	500*	~ 25	~ 20:1
Strain at failure ε_t	140 mm/m	0.06 mm/m	~ 2000:1

Characteristics of steel compared to those of concrete in tension (*yield strength)

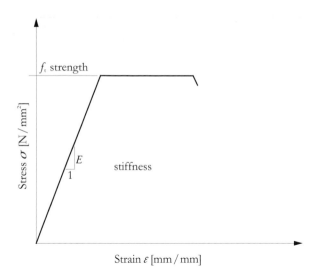

Mechanical behavior of the material

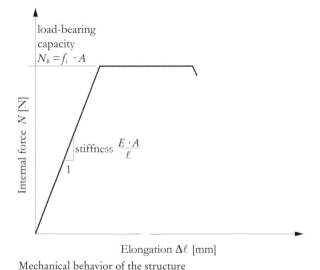

Mechanical behavior of the structure

The load-bearing capacity of the structure, on the other hand, corresponds to the strength of the material. The tensile or compressive internal force which, when reached, leads to the failure of the structure, can be easily calculated by multiplying the strength of the material by the area of the section:

$$N_R = f_t \cdot A \quad \text{in the case of failure by tension or}$$
$$N_R = -f_c \cdot A \quad \text{in the case of failure by compression}$$

It should be noted that the load-bearing capacity of the structures we have analyzed thus far does not depend on the form of the section or on the length of the structure. As we will see further on, this is true of structures subjected to tension and to compression when they are not excessively slender. When the slenderness (determined by the ratio between length and dimension of the section) goes beyond a certain limit, the collapse of the structure does not happen through failure of the material, but due to a phenomenon of instability. We will return to this problem at the end of the book. Before continuing, we should underline the two main characteristics of any material and any structure:

– *stiffness* indicates *how much* a material is deformed by internal force, or a structure by load: the greater the stiffness, the less deformation;
– *strength* indicates *how much* internal force can be imposed to a material or *how much* it can be loaded before failure.

Unfortunately, in common usage these terms are often confused. This is the result of the fact that both stiffness and strength depend on the type of material and the dimensions of the structure. Often, a stiff structure is also strong. So we need to pay attention to this distinction and to use the correct terminology, because the number of exceptions is infinite.

Dimensioning

The relationships that connect the geometry of the structure and the characteristics of the material to stiffness and strength allow us to determine the dimensions of the area of the section, in order to

– limit deformations
– avoid failure

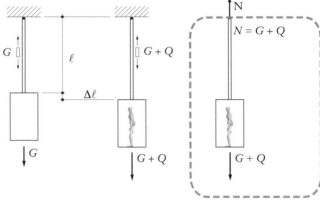

of the structure subjected to the loads envisioned. To better grasp this procedure, called *dimensioning*, we can consider the example of an elevator whose cabin is suspended on cables.

Criterion of the serviceability limit state (SLS)

The first condition makes it possible to determine the required area of the section of the cables so that the elongation $\Delta\ell$ caused by the weight of persons in the elevator will not exceed a certain limit $\Delta\ell_{all.}$ (admissible elongation), beyond which use of the elevator would become problematic, causing people to trip when exiting or giving rise to shaking that could make users feel unsafe (just imagine being in an elevator cabin suspended from an rubber band!).
The elongation $\Delta\ell$ caused by the variation of the internal force ΔN (corresponding to the load of the passengers Q) is:

G = weight of the cabin itself
Q = live load
N = internal force of the cable

The cabin of an elevator with the suspension cables; lowering of the cabin when loaded

$$\Delta\ell = \Delta N \frac{\ell}{E \cdot A}$$

and must be less than the allowable elongation $\Delta\ell_{all.}$. With this equation we can immediately determine the area of the section required to guarantee sufficient stiffness of the structure:

$$A_{req} \geq \frac{\Delta N}{E} \cdot \frac{\ell}{\Delta\ell_{all.}}$$

Imagine, for example, an elevator that can be loaded with a mass of 800 kg ($Q = \Delta N = 10 \times 800 = 8000$ N), with length of the cables equal to 30 m, a modulus of elasticity $E = 160\,000$ N/mm^2 and an allowable elongation of 10 mm (to avoid tripping). In this case we obtain the following condition:

$$A_{req} \geq \frac{8000 \text{ N}}{160\,000 \text{ N}/\text{mm}^2} \cdot \frac{30\,000 \text{ mm}}{10 \text{ mm}} = 150 \text{ mm}^2$$

Therefore we will have to use a cable with a diameter of at least 15 mm ($A = 177$ mm^2).

The limit value $\Delta \ell_{all.}$ can be determined according to the type of elevator and its planned use. This design criterion, then, refers to the state of "service" of the structure. This is why it is called the *criterion of the "serviceability limit state" (SLS)*.

Criterion of the ultimate limit state (ULS)

As we have seen, the other criterion serves to dimension the structure in order to avoid failure (in our case, that of the cables). This is an imperative criterion, because failure must be avoided at all costs (in real life, at all reasonable costs). For example, we will need to take all the most unfavorable cases of utilization into account. Although the admissible load is always clearly specified (max 8 persons, or max 800 kg), we know that this limit may be exceeded. Failure must clearly be avoided in this case as well.

Load factors

For this reason, an increased load is considered in the dimensioning, obtained by multiplying the expected load by a so-called load factor γ_Q. Actually, the elevator itself may be slightly heavier than expected. For this reason, its weight is calculated by multiplying its specified weight by a factor γ_G. Load factors are set by technical standards. The adjacent table shows indicative values found in European regulations for buildings and bridges. Other regulations provide similar values.

Dead loads	$\gamma_G = 1.35$
Live loads Wind pressure	$\gamma_Q = 1.5$

Load factors for calculation of factored loads (according to European codes)

Factored loads and design value of the internal force

Therefore dimensioning of the structure, will be carried out by considering the sum of all the possible factored loads. The internal force obtained by combining these loads is the so-called *design value of the internal force*:

$$N_d = \gamma_G \cdot G + \gamma_Q \cdot Q$$

Resistance factors

To be sufficiently prudent, we should also consider the fact that the strength of the material and the area of the section might be slightly less than planned. In fact, when a structure is designed, the exact characteristics of the material that will be used in the construction are not yet completely known. For this reason the load-bearing capacity of the structure must be reduced with respect to the expected capacity, dividing the declared strength of the material by another safety factor, depending on the type of material chosen.

In this way, we can take various levels of strength variation into account. Steel, thanks to its constantly controlled production, has much less strength variation than wood (which often has knots, shrinkage cracks and other possible imperfections), concrete and masonry.

The values in the adjacent table refer to construction materials. For structures subject to wear, as in the case of the cables of an elevator, much higher factors must be utilized. It must be noted that this approach is primarily used in Europe. In other parts of the world (North America for instance), a similar result is obtained by multiplying the minimal strength of the material by a "strength factor" lower than unity.

steel	$\gamma_M = 1.05\text{-}1.10$
reinforcing steel	$\gamma_M = 1.15$
concrete	$\gamma_M = 1.5$
wood	$\gamma_M = 1.5 - 1.7$
masonry	$\gamma_M = 2.0$

Safety factors used to calculate the reduced strength of materials (according to European codes)

Design strength

With the design strength of the material $f_d = f/\gamma_M$, we can obtain the design strength of the structural element:

$$N_{Rd} = f_d \cdot A$$

Comparing the latter with the design value of the internal force, which must always be less than ($N_d \leq N_{Rd}$), we obtain the required area of the section, dividing the design internal force N_d by the design strength f_d:

$$A_{req} \geq \frac{|N_d|}{f_d}$$

This design criterion is called the *criterion of the ultimate limit state* (ULS).

In the case of our elevator, assuming a weight of the cabin $G = 9000$ N (mass of 900 kg), we will have a design internal force $N_d = 1.35 \times 9000 + 1.5 \times 8000 = 24.150$ N.

With a material design strength of $f_y = 1410$ N/mm² and a resistance factor $\gamma_M = 6$ (caused by the danger of wear) we obtain a design strength of

$$f_d = \frac{1410}{6} = 235 \text{ N} / \text{mm}^2$$

and the required area of cable will be

$$A_{req} = \frac{24150}{235} = 103 \text{ mm}^2$$

In this case, the criterion of the serviceability limit state ($A_{req} = 150$ mm²) is decisive, not the criterion of the ultimate limit state. If, on the other hand, we permit a lowering of the cabin equal to 20 mm, the criterion of the ultimate limit state will become decisive, and in this case a cable with a diameter of 12 mm will suffice (A = 113 mm² > A_{req}).

Fatigue

When the material of a structure is subjected to internal forces that frequently vary over time, fragile failure can occur, even for internal forces well below the strength of the material.

This phenomenon, known as *fatigue*, is often decisive for railroad bridges, cranes and machines with rapidly moving parts. For architectural structures, on the other hand, fatigue is almost never a decisive factor.

Equilibrium of more than two forces in a plane and in space

The structures and examples we have examined thus far are marked by the action of all the forces and all the internal forces along a single line of action.

Now let us consider an example similar to the first one, with the person standing on the floor, but with an added force. Imagine the same person who remains in equilibrium by gripping a cable attached to the wall. To simplify things, assume that the cable is horizontal.

From experience, we know that in this situation the cable will be taut. In other words, it will be subjected to tension. This means that if we isolate a free body comprising the person and part of the cable, besides the gravitational force exerted by the earth on the person (Q = 700 N) and the thrust the floor exerts on the person corresponding to the surface of contact between the foot and the floor (R), we will have to consider the force the cable exerts on the free bodies (H). This force corresponds, clearly, to the internal force present in the cable. Since the cable pulls the free bodies, in this case the force we have to introduce will be directed from right to left.

If we analyze the force the floor exerts on the person, we notice that it can no longer be the usual vertical, upward force. This force will be inclined. In other words, as well as the vertical component there will also be a horizontal component.

The presence of this horizontal component can be demonstrated by thinking about what would happen were it to be removed, as in the case of a very slippery floor, or the case of a person on a trolley. In such cases, on gripping the cable and leaning back, the person would slide to the floor and equilibrium would no longer be possible.

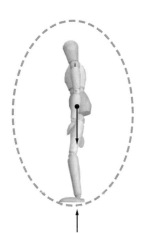

Subsystem and acting forces with the person standing on the floor

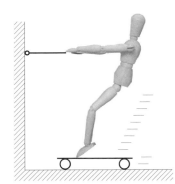

Person gripping a cable

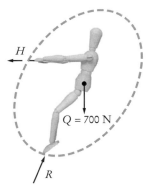

Subsystem and acting forces

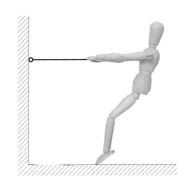

Case without the horizontal component of the action of the ground

First condition of equilibrium	Returning to the case of a sufficiently rough floor, we can find the first condition of equilibrium by extending what we have discovered previously regarding two forces that operate along the same line of action:

The forces that act on a free body are in equilibrium if they vectorially cancel each other out. |
| **Polygon of forces or force polygon** | We can represent this condition with the so-called *polygon of forces*. The forces cancel each other out to form the vector that closes the polygon. In other words, by adding one vector after another we return to the starting point.

The intensity of the gravitational force is known to us, because we know the mass of the person. The same cannot be said of the forces exerted by the floor and the cable.

To calculate these intensities, we must be able to represent the polygon of forces with the correct inclination of the force R exerted by the floor. |
| **Second condition of equilibrium** | We need a second equilibrium condition to determine the inclination. For the person standing on the floor, with only two forces, we had stated that these two forces must be on the same line of action. Here, with three forces, this condition can be reformulated as:

The lines of action of three forces in equilibrium must meet at a single point in the adjacent figure. |
| **Point of application of a force and equilibrium** | Note that this point does not necessarily correspond to the center of gravity. If the person lowers himself without varying the height of the cable, the position of the feet or the line of action of Q, she will not modify the situation of equilibrium at all. In other words, the point of application of a force can be shifted along its line of action without altering the equilibrium.

If, instead, we shift the position of the feet, we have a different situation of equilibrium, because the inclination of the force R exerted by the floor has been modified. Moving the feet towards the wall, as the polygon of forces shows, will cause an increase in the internal force in the cable. |

$Q = 700\ N$

Polygon of forces

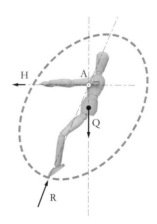

Exact solution

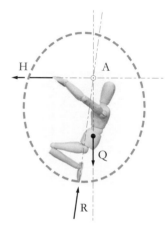

Case with lowered center of gravity; the condition of equilibrium remains unchanged

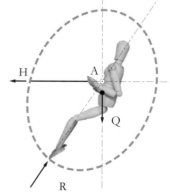

the internal force H in the cable increases

The variation of internal forces caused by shifting the position of the feet

Angle of friction

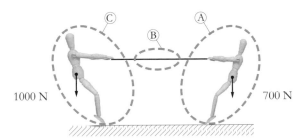

Two persons of different mass in a tug-of-war

1000 N

700 N

Experience teaches us that if we move our feet beyond a certain limit, even with a good grip on the floor, our shoes will slip and we will fall. This is due to insufficient friction between the floor and the soles of the shoes.

The limit is reached when the angle between the force exerted by the floor (or the force exerted on the floor) and a straight line perpendicular to the surface of the floor exceeds the so-called *angle of friction*. This limit angle depends on the roughness of the surfaces and the type of material. For example, a rubber sole on a concrete floor has a much larger angle of friction than a leather sole on an icy surface. In the first case, the person can shift her feet further forward, further inclining the force R and increasing the internal force in the cable. In the second case, on the other hand, to avoid slipping the person must remain in an almost vertical position.

Now we will analyze a very similar case, that of two persons in a tug-of-war.

Assume they have unequal masses: 70 kg for one, 100 kg for the other.

If we isolate "free body A" with the person weighing 70 kg, we notice that the forces involved are identical to those in the previous example. There is evidently no difference if it is the wall or the 100 kg person that exerts the force on "free body A" by means of the cable.

For each free body, we can trace a force polygon.

The force H that appears in all the free bodies must clearly always have the same intensity. In other words, the force the 70-kg person exerts on the cable and on the other person must correspond to the force exerted by the 100-kg person on the cable and, respectively, on the 70-kg person.

In the construction of the force polygon, to avoid repeating the inclusion of the same force we can combine multiple polygons in a single drawing, known as a *Cremona diagram*. This construction makes it even clearer that the two persons keep each other in equilibrium simply by varying the slope of the thrust exerted on them by the ground.
Note that the force exerted by each of the two athletes on the cable and, through it, on his adversary, depends only:
  – on his own weight;
  – on the capacity to incline the force exerted on the ground (in other words, on the angle of friction between the shoe and the ground).

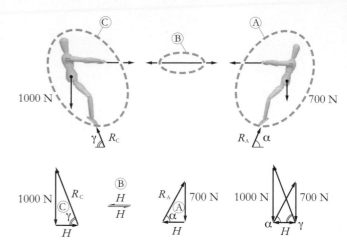

Subsystems and Cremona diagram

Forces and internal forces

The so-called "physical force" of the athlete, on the other hand, has no influence. Actually his muscles, together with bones and tendons, must simply be capable of transmitting internal forces through the body. As is shown in the illustration, the internal forces of compression and tension are on lines of action that can emerge from the body. Later we will see how these internal forces can be carried by the "load-bearing structure" of our body. For the moment it will suffice to understand that beyond the forces that act on the free body, we can also consider and analyze the internal forces inside the free body, shown as rods under compression and tension.

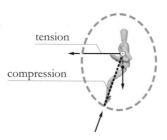

Resultants of the internal forces on the body

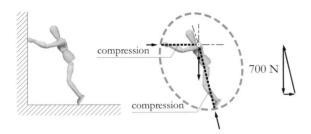

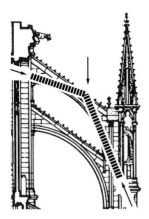

Person leaning against a wall, subsystem and force polygon

700 N

Similarity to the rampant arch of a Gothic cathedral

Let us complete this overview with another, very similar example and an actual structure that exhibits a similar behavior. Imagine that the person who was gripping the cable is now leaning against the wall. We know, by experience, that if her feet slide on the floor, unlike the previous example, they will move further away from the wall. This means that again in this case the force the floor exerts on the person is inclined, but in the opposite direction. If we isolate the "person free body" we see that to restore equilibrium we also need a force exerted by the wall, which in this case no longer pulls the person by means of the cable, but pushes him, exerting a force from left to right. This means that not only the legs but also the arms are subjected to internal forces, essentially by compression.

The examples we have analyzed make it possible to understand the conditions that must exist for a system, or a part of it (free body), to be in equilibrium. At the same time, however, we have already introduced some real structures.

Later we will see that the action of the person leaning on the wall is very similar to that of a rampant arch in a Gothic cathedral.

Cables

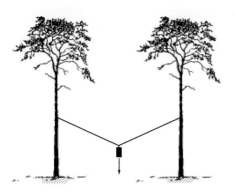

A cable hung between two trees can carry a load

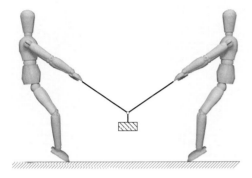

The athletes in a tug-of-war can support a weight

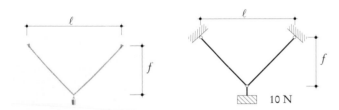

Cable model and corresponding structural diagram

First, we need to understand the functioning of structures under subjected to tension. With a cable we can support a weight, seen as a gravitational force, or, more generally, we can transmit a force whose line of action corresponds to the axis of the cable.

If we consider the case of laundry hung out to dry on a clothesline, or the weight of the automobiles supported by the cables of a suspension bridge, or other similar examples in architecture where roofing is supported by a system of cables, we see that a cable does not necessarily have to act in the line of action of the loads. In practice, this is the situation if we hang a weight halfway along a cable that two athletes are pulling in a tug-of-war.

Let us examine the functioning of a structure of this type by using a very simple example shown in the model in the illustration. A chain with its two ends attached to two points along the same horizon line supports a mass of 1 kg situated halfway along the length of the chain. By isolating a free body that includes only the mass, we can observe a gravitational force equal to 10 N exerted by the earth and a supporting force of equal intensity exerted by the structure. Therefore we can state that the mass exerts a load of 10 N on our structure composed of the cable and the supports.

Structural diagrams

Structures are often represented by means of a simplified diagram. The cable is replaced by two lines and the supports can be represented as seen in the structural diagram beside the text. As we will see, it is important to represent the structure to scale. In our case, the slope of the two cable segments should correspond to reality.

Experience tells us that under the influence of a load the cable tends to lower, until it reaches its position of equilibrium. The form the cable assumes is therefore that of a triangle. In a real situation this is true only if we assume that the weight of the cable itself is negligible with respect to that of the load it carries. We will see later what form the cable assumes if its weight is not negligible with respect to the load.

The span ℓ and the rise f

The geometry of this structure can be defined by the distance between the points of attachment (supports) and the difference in height between the supports points of attachment and the point of application of the load. In the following, we will use the terms span ℓ to define the horizontal distance between the attachment points, and rise f for the height of the structure.

To understand the functioning of this structure we can isolate a free body that includes the suspended weight and two segments of cable. Where the limit of the subsystem cuts the cables we have to insert internal forces in the cables. These forces exerted by the cables on the free body must have the axis of the cable as their line of action, so if the geometry is known we can outline the force polygon, to guarantee equilibrium. In other words, the three acting forces must cancel each other out vectorially.

The intensities of the internal forces in cables N_1 and N_2 can thus be easily found by measuring the length of the two vectors we have traced. Due to the symmetry of the system, the two intensities are clearly identical.

The second condition of equilibrium, which states that the three lines of action must meet in one point, in this case implies that the two cable segments must meet on the line of action of the gravitational force.

Supports

Now we can analyze the internal forces in the zone of the attachment points, also known as supports. By isolating the proper free bodies, we can analyze the force the cable exerts on the supports or, vice versa, the force the supports exert on the cable.

In particular, these forces can be broken down into a horizontal component H and a vertical component R_v.

Direction of the internal force on the subsystem

Note that the internal force N_1, as it is a tensile force, pulls the free body A upward, while at the same time pulling free body B downward. To indicate the action of the internal force in the correct direction we must always consider the type of internal force, tension or compression; in other words, whether it "pulls" or "pushes" a free body. Considering the fact that for the moment we are dealing above all with tensile internal forces, where we identify tensile elements with the free body, we will have to introduce forces that "pull". In other words, the vectors move away from the free body.

Because the internal forces we have identified take part in multiple free bodies, in this case it is also useful to combine the three force polygons in the so-called Cremona diagram.

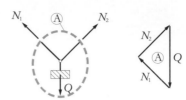

Free body including the suspended weight and the two segments of cable; force polygon

Free bodies of the supports, and force polygons

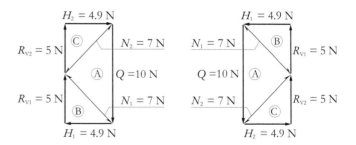

Cremona diagram composed of the three force polygons, variations with forces arranged clockwise and counterclockwise (the internal forces are rounded to a precision of 0.1 N)

The force polygon changes, depending on the order in which we examine the forces

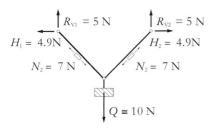

Type of internal forces, intensities and forces at the supports

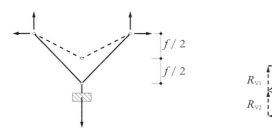

Influence of the rise on the internal forces

It should be noted that for any free body the force polygon changes depending on the order in which we consider the forces. For example, for free body B we can begin with N_1, proceed to H_1 and terminate with R_{v1}; or we can begin with N_1 but then move on to R_{v1} and H_1, in that order. To facilitate the composition of the Cremona diagram it is useful to always apply the same order, for example always consider the forces in the order determined by moving clockwise around the free body. We can obtain a different but equally correct Cremona diagram by moving counter-clockwise around the free body.

The Cremona diagrams make it clear that, due to symmetry, the intensity of the forces on the supports is equal ($H_1 = H_2$, $R_{v1} = R_{v2}$).
In particular, half the load is transferred to support 1 (R_{v1}), while the other half is transferred to the other support.

To conclude our analysis of the system, we can represent the type of internal forces, the intensities and the forces on the supports.
Note that in this structure composed of cables all the elements are subjected to tension.

Influence of the load

As can easily be imagined, the internal forces and the forces at the supports are directly influenced by the intensity of the load. If it is doubled, the internal and support forces will also double.

Influence of geometry

To understand the influence of geometry first we can halve the rise f while keeping the span ℓ the same. The two cables will have smaller slopes, so the force polygons of the three subsystems will also be modified.
Comparing the Cremona diagram with the previous diagram of the cable, we see that the vertical forces at the supports have remained the same. This is not surprising, because these vertical components correspond to half the vertical load, and the latter has remained unchanged. The internal forces in the cables have increased, while the horizontal components of the forces at the supports have actually doubled.
If we had doubled not only the rise but also the span, we would have a cable with the same shape as the original cable, but on a scale of 2:1. The slopes would remain the same, and therefore the Cremona diagram, the intensity of the forces and the internal forces would not undergo any variation.

The ℓ/f ratio

So it is clear that the internal forces and the forces depend only on the ℓ/f ratio and the intensity of the load.

As we have seen, the vertical component of the forces at the supports corresponds to half the load: $R_{v1} = R_{v2} = Q/2$.

To find the law that expresses the horizontal component of the force at the supports based on the load, of the span ℓ and the rise f, we can examine the affinity between the triangle of forces based on free body B and the triangle formed by the segment of cable involved, the rise f and half the span (ℓ/2).

In fact, H_1 is to ℓ/2 as Q/2 is to f. Expressed as an equation, the result is:

$$\frac{H_1}{\ell/2} = \frac{Q/2}{f} \quad \text{and therefore} \quad H_1 = \frac{Q \cdot \ell}{4 \cdot f}$$

The internal force N_1 can be found using Pythagoras' theorem:

$$N_1 = \sqrt{\frac{Q^2}{4} + \frac{Q^2}{16} \cdot \frac{\ell^2}{f^2}} = \frac{Q}{2} \cdot \sqrt{1 + \left(\frac{\ell}{2f}\right)^2}$$

For the forces on the other support and the internal forces N_2, the same equations are clearly valid, considering the symmetry of the system and its loads.

Influence of the position of the load

Now let us look at an example where the load is not hung at the center of the cable, but one quarter of the way along the span. With respect to the example analyzed above, the load, span and rise remain the same. Repeating the same procedure we used to analyze the symmetrical system, we can easily resolve this example as well.

As the Cremona diagram shows, the internal forces of the cables and the forces at the supports have changed.

The load is transferred above all onto the nearest support. The vertical component of the force at this support is equal, in fact, to three-quarters of the load, while the other force, corresponding to the remaining quarter, is transferred to the other support.

The horizontal components of the forces at the supports, on the other hand, are equal. This is true, in general, when the loads are vertical.

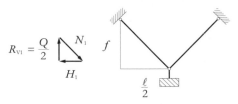

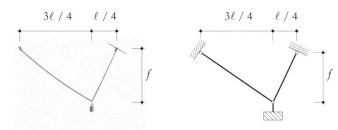

Affinity between the triangle of forces and the triangle determined by the segment of cable involved

Load hung one quarter of the way along the cable

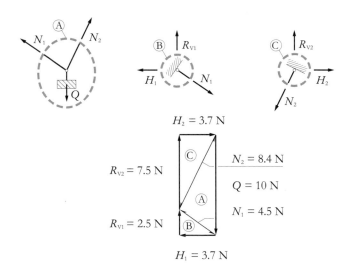

Free bodies and Cremona diagram (the internal forces determined are indicated with a precision of 0.1 N)

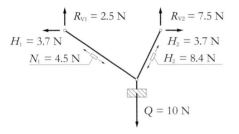

$R_{v1} = 2.5$ N $R_{v2} = 7.5$ N

$H_1 = 3.7$ N $H_2 = 3.7$ N

$N_1 = 4.5$ N $H_2 = 8.4$ N

$Q = 10$ N

Type of internal forces, intensities and forces at the supports

With respect to the symmetrical case, the internal force in the cable segment that connects the load to the nearest support has increased, while that of the other segment has decreased.

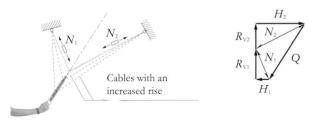

Cables with an increased rise

Load in any direction and Cremona diagram

Load in any direction

The case with a load that acts in any direction can also be approached in the same way. As we have seen, the condition according to which three forces are in balance if their three lines of action meet at a point requires the two cable segments to meet on the line of action of the load.

By varying the rise, we obtain the geometries of the cable shown in the illustration.

As the Cremona diagram shows, unlike the cases with vertical loads, the two horizontal components of the forces on the supports do not necessarily have the same intensity.

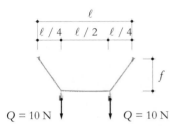

ℓ

$\ell / 4$ $\ell / 2$ $\ell / 4$

f

$Q = 10$ N $Q = 10$ N

Case with two vertical forces that act at ¼ and ¾ of the span

Cable with two vertical loads

Let us return to a case with vertical loads, but this time with two forces that act at one quarter and three quarters of the span. Under the action of the two loads, the cable will take on a polygonal form composed of three segments.

If the two loads have the same intensity and the two supports are at the same level, everything will be symmetrical. The central portion of the cable, then, can only be horizontal.

The structure can be analyzed by means of the free bodies A, B, C and D shown in the illustration.

The Cremona diagram shows that each of the two loads is transmitted only to the nearest support, while the tension force N_2, in the central horizontal portion, corresponds to the horizontal component of the forces at the supports.

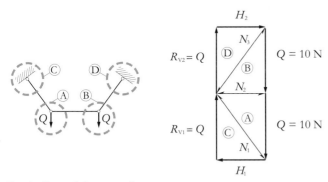

Free bodies and Cremona diagram

Note that the Cremona diagram is very similar to the one obtained for the first case we analyzed, with a single load acting halfway along the span. With a single load equal to $2Q$ and a rise taken to $2f$, giving the same slope of the cable portions near the supports, we will also have equal internal forces and identical forces at the supports. This situation is shown in the illustration.

If in the example with two loads we analyze a free body E including the whole central portion of the cable and both the loads, the force polygon will be identical to that of free body A in the example with a single, doubled load.

In the analysis of free body E, the two loads Q can clearly be replaced by their *resultant R*, seen as the vectorial sum. The line of action of the resulting force must reach the meeting point of the lines of action of the internal forces N_1 and N_2, as in the case with a single force. Here the line of action of the resultant clearly corresponds to the axis of symmetry of the system and the loads.

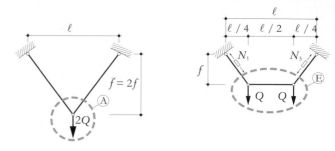

Similarity between the case with two forces and the case with a single load corresponding to the resultant force

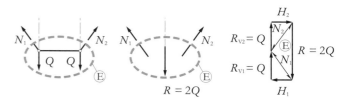

Equivalence of the two free bodies

Resultant cable

This construction using the resultant can be very useful for the study of the general case of two non-vertical forces with arbitrary geometry.

Cable with two non-vertical loads

The first step is to determine the resultant of the two loads: this is done by calculating their vectorial sum. Naturally the line of action of the resultant must meet with that of the loads at one point.

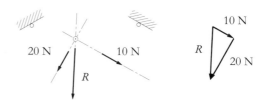

General case of two non-vertical forces with arbitrary geometry, determination of the resultant force

Once the intensity and line of action of the resultant R have been determined, it is easy to construct a cable capable of picking it up and transmitting it to the supports. As in the other cases analyzed, we can choose the rise $\bar{f}$ or the slope of one of the two portions of the cable. Note that, in general, the rise $\bar{f}$ of the cable capable of carrying the resultant does not correspond to the effective rise of the cable. In the case examined previously with two vertical loads in a symmetrical position, the rise of the cable that corresponds to the resultant was double the effective rise of the cable. Returning to the example with two non-vertical loads, we can observe that near the supports the resultant cable corresponds, again in this case, to the one that would be obtained with two real loads. The internal forces N_1 and N_3 can therefore be determined with the help of free body E. This construction applies starting from the supports all the way to the meeting of the lines of action of the two loads, where the cable will be deviated. So it is only in the zone located between the two

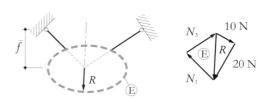

Resultant cable and analysis of the free body including the resultant

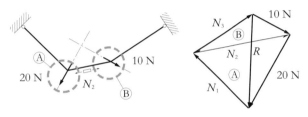

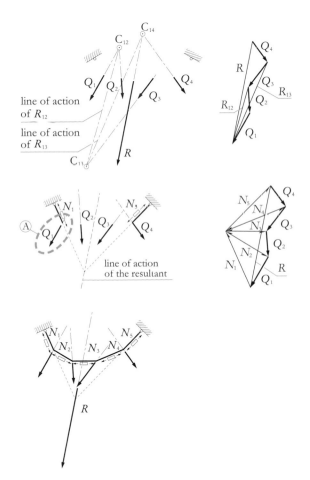

Real cable and analysis of the free bodies, Cremona diagram

General case and resolution procedure for finding the shape that the cable must assume and the internal forces

loads that the two cables, the real one and the one capable of carrying the resultant, are separated.

The slope the real cable will assume in this zone can be directly deduced from the Cremona diagram, analyzing free-body A.

If the construction is exact, the equilibrium of free body B will also automatically be satisfied. Note that this construction makes it possible to determine not only the internal forces, but also the shape that the cable must assume to be able to pick up the loads and transmit them to the supports.

This procedure can also be utilized when multiple forces are involved. In the particular case in which the lines of action of all the forces meet at one single point, the line of action of the resultant can immediately be determined. The direction can be determined by means of the Cremona diagram, where the line of action of the resultant can be traced by, making it pass through the point of intersection of the lines of action of the loads.

In the general case with lines of action of the loads not meeting at a single point, on the other hand, we need to proceed in steps.

Cable with multiple non-vertical loads

1. Determine the partial resultant R_{12} of the first two loads Q_1 and Q_2. Its line of action will pass through the point of intersection C_{12} of the lines of action of Q_1 and Q_2.
2. Determine the partial resultant R_{13} of the first three loads, adding Q_3 to R_{12}. Its line of action will pass through the point of intersection C_{13} of the lines of action R_{12} and Q_3.
3. Repeat this procedure, adding one force at a time until you have obtained the resultant R and its line of action.

At this point the problem is similar to the case with two forces. We can identify a possible cable capable of picking up the resultant, choosing the rise $\tilde{f}$ corresponding to the resultant, or the slope of one of the two portions of cable connected to the supports. By comparison with the first (free body A) or the last force, we can then analyze how the cable must be deviated, and repeating the operation force by force we can finally obtain the complete form of the cable.

Parallel non-symmetrical loads

In principle, the procedure can also be used when all loads are vertical. The intensity of the resultant can immediately be determined by summing the loads. Where its line of action is concerned, if the case is not symmetrical, the solution is not immediate. Here too, the line of action of the resultant meets those of the loads at one point, but *at infinity*, because all the lines are parallel!

Since the resultant and its line of action depend only on the loads, and are not influenced by the type of cable or its supports, the solution can be found by using the reverse of the method described above.

Auxiliary cable

In practice, we can start with the Cremona diagram by choosing the slope of the first cable portion and its internal force. After having made this choice, we can then find the slopes of the other cable portions, still on the basis of the Cremona diagram. Returning to the geometric situation, we finally can trace a cable that meets the conditions of equilibrium, but not necessarily the geometric conditions at the supports. For this reason, we will call this construction the *auxiliary cable*. It allows us to determine a point on the line of action of the resultant by simply extending the first and last segments of the cable until they meet.

Note that the initial choice of the slope of the first cable portion and its internal force has nothing to do with the choice of the project. We simply want to choose a cable whose sole purpose is to permit the determination of the line of action of the resultant; it does not necessarily have to coincide with the real cable.

Once we have found the line of action of the resultant, we can forget about the auxiliary cable. At this point the procedure is identical to the one applied in the previous cases: the choice of a real cable corresponding to the resultant able to fulfill the geometric conditions at the supports, construction of the second portion of cable, considering the deviation of the first load, construction of the other remaining portions, finishing with the determination of the internal forces and the forces at the supports.

The choice of the real cable, then, is part of the project. It is possible to choose cables with a smaller rise and therefore, higher internal forces, or cables with a larger rise, to reduce the internal forces.

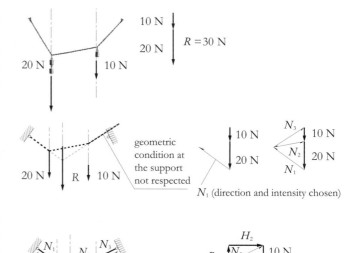

Case with two parallel non-symmetrical loads and resolution procedure using the auxiliary cable (the dotted line indicates the auxiliary cable, the continuous line represents the real cable)

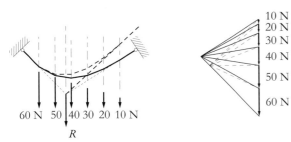

Example with six vertical loads

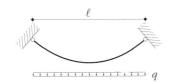

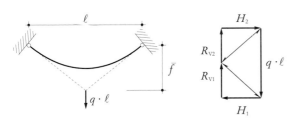

Cable subjected to uniformly distributed loads, the resultant cable and Cremona diagram

In the example illustrated here, the slope of the first portion of the real cable corresponds to that of the auxiliary cable. But we should remember that this coincidence is by no means indispensable.

Center of gravity

The determination of the resultant force by means of an auxiliary cable can also be used to find the *centre of gravity of a body*. All we need to do is to subdivide the body into parts and apply the corresponding gravitational force. The resultant of these forces will necessarily pass through the center of gravity. If knowledge of the intensity and its line of action is not enough, and we also want to determine its point of application (namely the center of gravity), we need to repeat the operation with fictional non-vertical forces and find the meeting point of the two lines of action thus constructed.

Funicular polygon

In the examples we have examined thus far, the cables take on configurations of equilibrium, marked by straight lines between each load and changes of direction occuring when the cable meets the lines of action of the loads. The polygon, then, is defined by the loads themselves, by the position of the supports and the rise, and can be constructed with the methods described above. This geometric figure is called the *funicular polygon*.

Distributed loads

Often, in architecture, loads are distributed instead of concentrated. Just consider, for example, the weight of the structures themselves or the weight of snow. These distributed loads can be seen as the sum of infinitesimal concentrated loads, positioned next to each other. In these cases the funicular polygon is then composed of infinite segments of infinitesimal length, and is thus a curve: so we can talk about a *funicular curve*.

Cable subjected to uniformly distributed loads

Let us consider the example of a cable subjected to a distributed load of constant intensity. To distinguish it from concentrated loads, for which capital letters are used (Q), distributed loads are indicated with lower-case letters (q). The intensity of the load is defined as the force acting on a unit of length and expressed in kN/m or N/m.

The resultant of the load that acts on the structure we are analyzing is therefore equivalent to $R = q \cdot \ell$, while its line of action must be at mid-span, due to the symmetry.

Therefore the funicular polygon of the resultant with rise $\bar{f}$ and the Cremona diagram with the resultant are elements we have already examined. We can get closer to the funicular curve if we approximate the distributed load by a series of forces, all of them equal, distributed at a constant distance from one another.

Imagine, for example, that we have subdivided the span into eight equal segments. The load that acts on each of these segments is equal to $q \cdot \ell/8$, and its line of action must pass through the halfway point of each segment. The Cremona diagram and the funicular polygon can then be constructed using the method already described.

Comparing the Cremona diagram we have just constructed to that of the resultant $R = q \cdot \ell$, and considering the equations already derived from the case of the cable with a concentrated load at its center, we can directly find certain relationships by replacing Q with $q \cdot \ell$.

The forces at the supports are

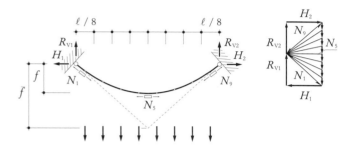

Span subdivided into eight equal segments, Cremona diagram and funicular polygon

$$R_{v1} = R_{v2} = \frac{q \cdot \ell}{2} \text{ and } H_1 = H_2 = \frac{q \cdot \ell^2}{4 \cdot \bar{f}}$$

The maximum internal forces in the cable correspond to those due to the resultant

$$N_1 = \frac{q \cdot \ell}{2} \cdot \sqrt{1 + \left(\frac{\ell}{2 \cdot \bar{f}}\right)^2}$$

while the minimum internal force in the central zone (N_5) has the same intensity as the horizontal forces H_1 and H_2 at the supports.

The funicular polygon constructed with eight forces has a rise f that is exactly half of that chosen for the resultant ($\bar{f}$). This will also be true if we further increase the number of forces in order to get closer to the funicular curve. Replacing $\bar{f} = 2 \cdot f$ in the earlier equations, we obtain expressions that are also valid for the case of a uniformly distributed load:

$$H = N_{\min} = \frac{q \cdot \ell^2}{8 \cdot f} \quad \text{and} \quad N_{\max} = \frac{q \cdot \ell}{2} \cdot \sqrt{1 + \left(\frac{\ell}{4 \cdot f}\right)^2}$$

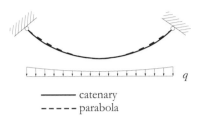

q

——— catenary
----- parabola

Difference between the geometric forms of a catenary and a parabola

Dulles Airport, Virginia, 1958-63, Arch. E. Saarinen, engineering Ammann & Whitney. (ℓ = 49 m, f = 8.25 m, ℓ/f = 5.94)

Golden Gate Bridge, California, 1937, Eng. J. Strauss
(ℓ = 1280 m, f = 160 m, ℓ/f = 8)

It can be demonstrated that the funicular curve with a distributed load of constant intensity is a second-degree parabola. For those interested in this demonstration, appendix 1 on page 231 shows the analysis of an infinitesimal element that permits us to derive the equation of the funicular curve.

Catenary

When the cables are subjected only to their own weight, we need to consider the fact that the weight is constant if measured along the line of the cable, but not horizontally, as in the case analyzed above. This means that where the cable has a greater slope, the weight per unit of horizontal length is larger than that found in the central part, where the slope is limited. In this case the cable assumes a shape different from a parabola.

The new form is called a *catenary* and can be described by the equation derived in appendix 1 on page 232. When the relationship between rise and span is not too large, so that the slope of the cable near the supports does not exceed a certain limit, the difference between the catenary and the parabola is minimal.

The illustration shows an example of a reinforced concrete roof. Because its own weight uniformly distributed along the length is dominant with respect to the other loads, the roof assumes the form of a catenary.

Suspension bridges

Often, the self-weight and the variable loads do not act directly on the cable, but are suspended by means of secondary cables called hangers from the load-bearing cable. This is the case with suspension bridges, in which the weight of the bridge surface, with its almost constant horizontal configuration, is much larger than that of the hangers and the main cable, with the result that in this case the geometric form assumed by the main cable is closer to a parabola.

The illustrations show the Golden Gate Bridge in San Francisco, with its construction phases. During installation of the main cables the form is that of catenary. On adding the weight of the bridge surface, the form approaches that of a parabola instead.

Clearly, the hangers transmit concentrated forces (on the main cable), so the form is, in reality, polygonal.

If we closely observe the main cables of the suspension bridge we can distinguish three parts that can be isolated into three subsystems.

In the central span the cable has a geometry similar to the one we have already analyzed with the supports positioned at the same height. The forces the cable transmits to the supports can be broken down. The vertical component will be taken up by the pylons, while the horizontal component will be balanced by the one found in the cables of the lateral spans. The latter have supports at different heights: one at the top of the pylons, the other at the foundation block, where the internal force of the cable is transmitted (horizontal and vertical component of the force at the supports) to the ground.

We can also think of the main cable as a single element that extends from one end to the other. In this case, the limit of the subsystem intersects:

– the two ends of the load-bearing cable, where the two support forces act relative to the anchoring blocks;
– all the hangers, where the loads act;
– and also the two pylons, under compression, meaning that we have to insert two internal forces that push upward.

Therefore the form assumed by the main cable over its entire length is the result of the funicular polygon obtained by considering the two supports at the ends and the downward loads, to which we must add the two upward thrusts transmitted by the two pylons.

Applications in architecture

Structures of this type also have their applications in architecture. The figures shown here are an example in which a roof is suspended by hangers from main cables. Again in this case, the roof is much heavier than the system of cables and the hangers are very close to one another, so the funicular polygon approaches a parabola. The main cables of the central span transmit their forces to the pylons, as in the case of the suspension bridge. In this case, the action of the lateral cables is different, as they no longer transmit their internal forces to two supports anchored to the ground, but to the roof itself. Later we will look at this example in greater depth, showing that the cable, combined with the roof under compression, forms a new type of structure.

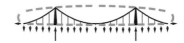

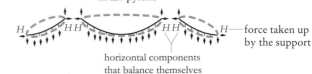

Suspension bridge, three subsystems, isolating the three segments of the main cable, and subsystem with the entire main cable and the pylons

Burgo paper mill, Mantua, Italy, 1960-64, Arch.-Eng. P.L. Nervi (ℓ = 160 m, f = 25 m, ℓ/f = 6.4)

In the examples we have just described the cables support, directly or indirectly through hangers, the roofs of buildings or the surfaces of bridges. In all the examples we can clearly see the importance of the elements indicated, up to this point, as *supports*. Actually, they are composed of columns, pylons, anchors in the ground or other parts of the support structure.

For the sizing of cables, we use the two criteria of the ultimate limit state and the serviceability limit state already described. The criterion according to which a structure must be dimensioned to make any failure highly improbable allows us to determine the section of the cable. In the case of cables composed of harmonic steel, the criterion of the ultimate limit state is completed with another, more restrictive condition. To avoid exceeding by large irreversible deformations and to limit, at the same time, the problems that usually arise in the anchoring zone of the cables, the tension due to permanent loads should be limited to about $0.45 \cdot f_t$.

When the loads are vertical the largest internal forces are found where the cables have the steepest slope. This generally happens near the supports. For the dimensioning of the cables, then, it will suffice to consider the internal forces in these zones. As we have seen, these forces depend on the intensity and the position of the loads, and on the ratio ℓ/f. The influence of this ratio, called the *slenderness ratio*, is illustrated in the diagram, which represents the functions already derived for the case of a cable with a load concentrated halfway along its span, and with a load uniformly distributed throughout its length. Along the ordinate, the relationship between the maximum internal force in the cable and the sum of the acting loads is represented. The increase in internal force and, therefore, in the section of the cable, with the growth of the slenderness ratio ℓ/f is evident. When the slenderness is limited because the span is small with respect to the rise, the maximum internal force equals about half of the loads. When, instead, we have a slenderness ratio that reaches 15, the internal force corresponds to two, or up to four times the sum of the loads.

The increase is much more rapid in the case with a concentrated load. This is due to the fact that with a distributed load the rise of the resultant (a parameter that, as we have seen, is determined for the maximum internal force near the supports) is twice that of the real rise.

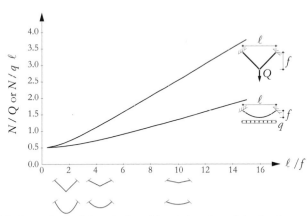
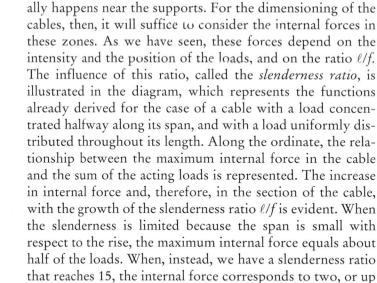

Maximum internal force in the cable as a function of the slenderness ratio ℓ/f

The total quantity of material utilized depends not only on the area of the dimensioned section, but also on the length of the cable. This latter factor also varies based on the slenderness ratio ℓ/f. As indicated in the diagram, the situation is inverted with respect to that of the internal force: when the ratio ℓ/f increases, the length L of the cable decreases.

Quantity of material based on the slenderness ratio ℓ/f

The two factors only partially compensate for each other. For limited slenderness the length prevails, while in the case of slender cables the increase of the internal force and, therefore, of the required section, becomes decisive. The diagram shows the increase in the quantity of material based on the slenderness ratio. The values plotted on the ordinate have been made dimensionless, dividing the quantity of material of the cable by the quantity of material that would be necessary to support the entire load Q or $q \cdot \ell$ by means of a vertical cable with length ℓ (see the example of the elevator on p. 26).

The curve that refers to the cable with a concentrated load has its minimum when $\ell/f = 2$. This means that the greatest efficiency is achieved when the slope of the two portions of the cable is equal to 45°. For the cable subjected to a distributed load, two curves are shown: the lower curve represents the infrequent case of a continuous variation of the section, in order to completely exploit the material, while for the upper curve the section subjected to the maximal internal force is maintained throughout the length of the cable. In the first case the greatest efficiency is achieved at $\ell/f = 2.31$, while in the second case the optimal slenderness ratio is 2.93.

Movements caused by variation of the intensity of loads

In the constructions described above, excessive deformations can interfere with their functioning. In the case of roofs, the movements must be limited so as not to cause damage to other structural and non-structural elements, especially those of the facade. Moreover, deformations beyond a certain limit could prevent drainage of water from the roof.

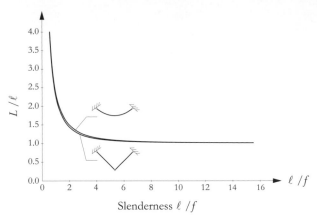

Length L of the cable in relation to the slenderness ratio ℓ/f (with the same span ℓ)

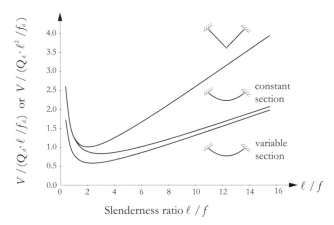

Amount of material as a function of the slenderness ratio ℓ/f (for a span ℓ, load $q_d \cdot \ell$ or Q_d and strength f_d)

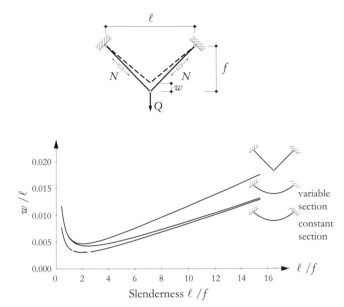

Mid-span deflection caused by concentrated or distributed loads in relation to the slenderness ratio ℓ/f ($\sigma = 0.45 \cdot f_t = 707$ N/mm², $E = 160\,000$ N/mm²)

As shown in the adjacent example, the increase in load intensity causes an increase in internal force, which, under conditions of elasticity, causes in turn a proportional lengthening of the cable. This deformation is reflected in a movement whose pattern is similar to the funicular polygon.

With normal construction materials, thanks to their high modulus of elasticity, unit deformations are very small. If we consider the use of a harmonic steel cable up to a tension of $0.45 \cdot f_t$, we will have a unit deformation of just 4.4 mm/m ($f_t = 1570$ N/mm², $E = 160\,000$ N/mm²).

In large structures, even relatively small unit deformations can cause excessive deflections. In the diagram shown here the movement at mid-span of a cable, with the degree of use of the material and lengthening already described, is shown in relation to the slenderness ratio ℓ/f.

In this case, too, for very limited slendernesses, or excessive slendernesses, the movements are large, while acceptable values are found in the intermediate range. For the real case of a parabolic cable with $\ell = 80.00$ m and $f = 8.00$ m ($\ell/f = 10$), the diagram shows a relationship of w/ℓ, equal to about 0.0087, meaning a movement $w = 0.0087 \cdot 80.00 = 0.70$ m! Such a movement would probably be unacceptable for any secondary structure.

We should remember that a major part of this deflection is caused by permanent loads that usually do not have to be considered in the verification of the limit state. These deflections, in fact, can be cancelled by installing cables that are slightly shorter than the theoretical length, to compensate for the elastic deformation caused by permanent loads. Moreover, the susceptible secondary structures are assembled after installation of the cables and the roof, when the deformations caused by permanent loads have already been almost completely established.

Deflections caused by permanent loads

Deflections caused by variable loads, whose effect will completely act on the secondary elements, can easily be calculated assuming a linear behavior of the structure. Considering the relationship between variable loads q and permanent loads g, we have: $w(q) = w(g+q) \cdot q / g+q$. If this movement is unacceptable and the span ℓ cannot be reduced, we must increase the rise f, in order to reduce the slenderness ratio (as shown in the diagram), or reduce the deformation of the cable by increasing its section. In this case we can use a steel of lower strength without compromising the criterion of the ultimate limit state.

Movements caused by variable loads

Deflections caused by temperature variation

We have seen that an increase in the load causes an elastic lengthening of the cable that results in a vertical movement. A similar situation arises with a temperature variation, which also causes a unit deformation of the material that can be quantified with the following formula:

$$\varepsilon = \frac{\Delta\ell}{\ell} = \alpha \cdot \Delta T$$

where ΔT is the temperature variation in °C et α is the constant of thermal expansion that depends on the material. In the case of steel, for example, $\alpha = 0.00001$ °C^{-1}. The diagram adjacent shows, in relation to the slenderness ratio, the pattern of vertical deflection caused by an increase in temperature equal to 20 °C ($\varepsilon = 0.0002 = 0.2$ mm/m). Note the resemblance to the previous diagram, proof of the similarity of the effects. We should observe that in this case the unit deformation does not depend on the section of the cable. Once the material has been chosen, the deflection caused by an increase in temperature can only be reduced by varying the geometric parameters ℓ and f.

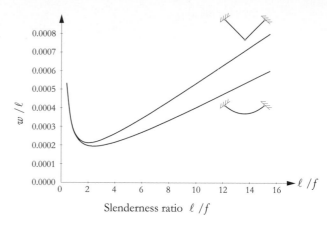

Slenderness ratio $\ell\,/f$

Vertical deflection caused by the increase in temperature $\Delta T = 20$ °C in relation to the slenderness ratio ℓ/f ($\alpha = 0.00001$ °C^{-1})

Effect of horizontal movements of the supports on the geometry of the cable

As shown in the illustration, a horizontal movement of the supports will also cause a vertical displacement of the cable. In fact, if the supports are brought closer together – in practice, a slight reduction of the span in relation to a constant length of the cable – the effect is similar to that of the lengthening of the cable with a constant span, as analyzed above.

Effect of horizontal displacements at the supports on the cable geometry.

Variation of the configuration of the loads

When both permanent and variable loads act on a cable, there can often be variation of the distribution of the load, leading to a change in the funicular polygon. This happens because the variable load, due to its changing conditions, can be distributed differently with respect to the permanent load.

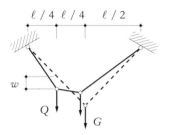

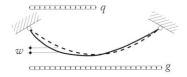

Variation of the configuration of the loads

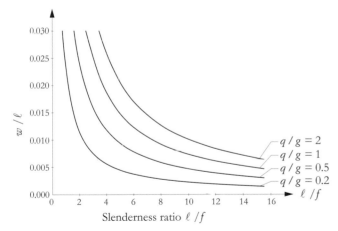

Displacement measured at one-quarter of the span caused by a variable load distributed on one half of the span, in relation to the slenderness ratio ℓ/f; parameter for the four curves: relationship q/g (neglecting the elongation of the cable)

The example in the illustration shows a cable subjected to the action of a permanent load G at mid-span, and a variable load Q acting at one quarter of the span. Without the variable load Q, the cable, subjected only to the permanent load G, assumes a symmetrical triangular shape. When the variable load Q is added, the cable takes on a polygonal shape composed of three segments. Varying the load Q, then, we get clearly visible deflection of the cable. Note that they are caused not only by the lengthening of the cable, but also and above all by the change of the funicular polygon and, therefore, of the shape of the cable.

Movements caused by variable loads

A similar situation arises when distributed loads are involved. Consider, for example, the case of a cable subjected to a permanent load g uniformly distributed across the span, and a variable load q composed of the weight of snow that, due to wind for example, can be concentrated on one half of the cable.

The diagram represents the displacement w in relation to the slenderness ratio ℓ/f for four variable load values. What is decisive is not the absolute value of the variable load q, but the relationship q/g between the variable load and the permanent load. Increasing this relationship, leads clearly to greater displacements. The diagram also provides a good illustration of the influence of the slenderness ratio ℓ/f. For cables with small slenderness the deflections are very large; by making them more slender, we can diminish this effect. If we examine our concrete case with $\ell = 80$ m and $f = 8$ m, we obtain a deflection of 0.58 m ($w/\ell = 0.0072$) when the variable load is equal to the permanent load, while it reaches 0.80 m ($w/\ell = 0.0100$) if the variable load is double with respect to the permanent load.
As we have already seen, these deflections are not the result of the deformation of the material, so even if we increase the area of the cable, we will not achieve any increase in rigidity.

As the examples described above illustrate, a simple cable used in architecture, in which other elements act together, will rarely be able to satisfy the serviceability limit state criterion. Therefore it is indispensable to intervene to limit the displacements derived from the change of form caused by variable loads.

Limiting displacements caused by variable loads

Cable subjected to permanent loads

Maison des Jeunes et de la Culture, Firminy-Vert, France, 1961-65, Arch. Le Corbusier (ℓ = 18.25 m, f = 1.30 m, ℓ/f = 14.04 m) [© 2004, FLC/ProLitteris]

Increase of the permanent load

As we have seen, for a given geometry, the displacement depends only on the ratio between the variable and permanent loads. Because is hard to influence the variable load (snow, wind, people on the roof), this ratio can be improved only by increasing the permanent load.

It is evident that with this solution we will also increase the internal forces in the cables and the structures of the support, with a corresponding increase in the amount of material. In spite of this drawback, the solution is applied at times, in particular for spans that are not very large. The illustrations show a roof designed by Le Corbusier in which panels of reinforced concrete rest on harmonic steel cables.

This solution can also resolve another problem with funicular structures. In fact, when the permanent load is too small, the suction of wind on the external surface or an increase of pressure inside the construction due to the wind can raise the roof.

To quantify the effect of the permanent load on the movements, we can return to our case with ℓ = 80 m and f = 8 m. Keeping the variable load constant and increasing the permanent load until we reach a relationship q/g = 0.2, we reduce the displacement to 0.18 m (w/ℓ = 0.0022, the lower curve in the diagram on the previous page).

Solution with stiffening cable: cable beams

One solution very similar to the previous one, at least in functional terms, is the application of permanent loads that are not gravitational forces (weight) but actions exerted by another cable. A stiffening cable with a downward curvature, linked by connection cables to the load-bearing cable, is capable of performing this function. The resulting system is called a *cable beam*. The diagram represents the internal forces induced by the stiffening of the system. Isolating the free bodies composed of the load-bearing cable, the stiffening cable and the connection cables, we can see the actions mutually exerted by the three subsystems, also without the presence of external loads. The forces P_i are introduced by horizontally moving a support of the load-bearing cable, shortening the connection cables, or moving a support of the stiffening cable.

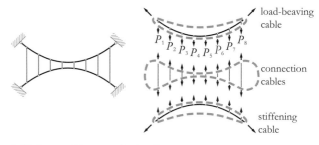

load-beaving cable

connection cables

stiffening cable

Cable beam with pretensioning forces

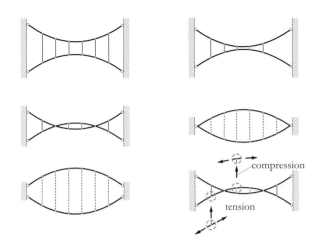

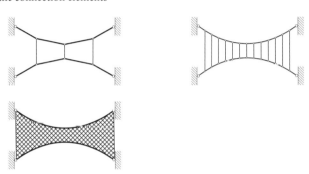

Five possible cable beams with different distances between the load-bearing cable and the stiffening cable; study of the internal forces in the connection elements

compression

tension

Cable beams with a few, many and infinite connection cables

Furthermore, this structural solution makes it possible to avoid the raising of the roof due to the effects of suction or pressure caused by wind.

Cable beams can assume a variety of shapes. In the design of these structures, for a given span, the parameters that can vary are: the rise of the load-bearing cable, that of the stiffening cable, the distance between the two cables and the geometry of the connection cables.

The figure show examples in which the distance between the load-bearing and the stiffening cables varies. Where the stiffening cable rises above the load-bearing cable, the connection elements are no longer in tension, but in compression. This particularity can easily be seen by isolating the subsystems, as shown. In these cases, the connection cables therefore have to be replaced by struts to resist compression.

Varying the intervals between the connection cables, we also obtain different geometries for the load-bearing and stiffening cables. When this distance becomes very small the form of the load-bearing cable approximates a curve. The connection cables can also be replaced by membranes: in this case the load-bearing cable and the stiffening cable effectively assume the form of a curve.

The connection cables do not necessarily have to be vertical. The cable beam represented here has diagonally arranged connection cables. This system, introduced and developed by the engineer David Jawerth in the 1950s, in characterized by its great efficiency and is very rigid under the influence of non-symmetrical loads. We will return to the functioning of diagonal connection cables later on.

Johannishov Ice Stadium in Stockholm, Sweden, 1962, Arch. Hedqvist, Eng. Jawerth ($\ell = 83$ m)

Cable beams, because they are composed mostly of elements under tension, are characterized by great lightness and, above all, by extreme transparency. This is why the engineer Peter Rice applied them in the 1980s as stiffening structures for large glazings or greenhouses, with the function of resisting wind thrust.

In spite of the great range of possible shapes, all cable beams have a load-bearing cable with a curvature directed toward the main loads (generally upward) and a stiffening cable curved in the opposite direction.

Glasshouses of the Parc Citroën, Paris, France 1992, Arch. P. Berger, Eng. P. Rice and D. Hutton

Glazing of the Cité des sciences et de l'industrie, Paris, France, 1986, Arch. Fainsilber, Eng. P. Rice and D. Hutton

Solution with load-bearing cable and stay cables

One solution that is very similar to the cable beam consists of a load-bearing cable stiffened by a system of cables directly attached to the lower supports (stay cables).

It works in a similar way to that of the previous solution: the stiffening elements, if pre-tensioned, exert a permanent load on the load-bearing cable. Furthermore, they can limit the raising of the load-bearing cable in the zone bearing less load. With a small upward shift of the load-bearing cable, in fact, a lengthening of the stay cables is caused, which thanks to their elasticity is transformed into an increase of the internal force corresponding to an additional load on the load-bearing cable, capable of stabilizing it.

decrease of internal forces   increase of internal forces

Pretensioning of the stay cables (without variable load)

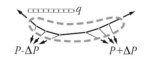

Increase of internal forces in the stay cables that limit the rising of the load-bearing cable in the zone with less variable load

Pavilion 26, Deutsche Messe Hannover, Germany, 1996, Arch. Herzog, Eng. Schlaich

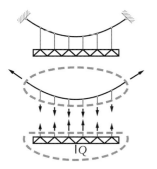

Load-bearing cable stiffened by a beam; transmission of a concentrated load by means of a parabolic cable

Tacoma Narrows Bridge, near Seattle, Washington, 1940, Eng. Moisseiff ($\ell = 853$ m, $f = 70.7$ m, $h_{deck} = 2.44$ m), collapse on November 7th 1940, caused by wind

In like manner, the stay cables tend to shorten when the load-bearing cable lowers under the effect of the variable load. As they do so, the force of pretensioning tends to diminish. This, however, represents a weak point of these structures: if the decrease in internal force is greater than the pretensioning force, the stay cable tends to slacken and be deactivated. In the example shown here, real springs have been introduced in the stay cables in order to resolve this problem.

The cable with stiffening beam

Another structural solution, introduced to limit the movements caused by the change of form due to variable loads, is that of the load-bearing cable linked by means of secondary cables to a stiffening beam. The beam, thanks to its stiffness, opposes the movements and, in doing so, redistributes the variable loads, making the load on the load-bearing cable, represented by the internal forces of the secondary cables, approach the load corresponding to the funicular curve. The diagram shows the case of a concentrated load halfway along the span. Without the stiffening beam, the shape of the cable would tend to approach that of a triangle, with a significant movement in the central zone. The stiffening beam, on the other hand, permits distribution of the load to multiple secondary cables. To do this the beam undergoes deformations that have repercussions for the geometry of the load-bearing cable and the internal forces that act between the two load-bearing elements.

This solution is often used for suspension bridges. The bridge surface, crossed by automobiles or trains, must be sufficiently stiff to limit the deformations caused by variable loads and the movements caused, above all, by the wind. In the case of the *Tacoma Narrows Bridge*, built with a stiffening beam that was too slender, the wind caused such large deformations that the bridge collapsed. After this accident, which could not have been predicted with the knowledge available at the time, builders returned to the use of more rigid bridge-beams. An evolution in the 1960s, using beams with an aerodynamic, boxed section, again permitted a reduction of the depth of the bridge-beams.

In architecture this type of structure is used to cover large spans. The structure of the roof acts, in this case, as the stiffening beam, while the parabolic cables remain exposed. The Burgo paper mill by Pier Luigi Nervi, which we have already seen, is a typical example (page 48). The illustration shows another case marked by two load-bearing cables, supported by two pylons and directly anchored to the ground.

Europahalle, Karlsruhe, Germany, 1983, Arch. Schmitt and Arch. Kasimir, Eng. Schlaich

Cables with flexural stiffness

The action is similar when the stiffening beam and the load-bearing cable are combined into a single element. Under the influence of permanent loads, the structure behaves like a cable, while the variable loads, both concentrated and distributed along a part of the length, are split in the same way as in the previous case. Thanks to its flexural stiffness, the cable does not have to move until it reaches the funicular polygon of the loads. This means that when there is flexural stiffness, the line of action of the internal forces does not necessarily coincide with the axis of the structure.

The structure can be composed by a system of interconnected cables, or more probably of beams or trusses bent or assembled with the form of a catenary.

If the structure is sufficiently rigid, a form can be chosen that does not correspond to the funicular curve of the permanent loads. This particularity permits easy adaptation for other needs.

Load-bearing cable with flexural stiffness

Tokyo Olympic Center, 1964, Japan, Arch. K. Tange, Eng. Tsuboi, M. Kawaguchi, S. Kawamata

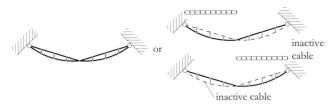

Combined cables for picking up variable loads

Tower Bridge in London, UK, 1894, Arch. H. Jones and G.D. Stevenson, Eng. J.W. Barry

Combination of multiple cables for picking up different loads; cable-stayed system

Maintenance center for Boeing 747s at Rome-Fiumicino, Italy, 1969-70, Eng. R. Morandi (ℓ = 80.00 m)

Systems with combined cables

Yet another solution for limiting the movements caused by variable loads consists in combining multiple cables in the same system to permit, depending on the position of the load, activation of the cable capable of taking up the load with the least movement. If distributed loads are expected, acting on one half of the structure or the other, two cables can be arranged as in the illustration.

This system was used for the lateral spans of Tower Bridge in London. In this case, the two cables are connected by diagonals that transform them into an element with flexural stiffness, as described above.

Cable-stayed systems

It is also possible to combine many cables with polygonal forms that can individually take up concentrated forces, but when combined can also easily resist distributed loads. Often these structures, known as *cable-stayed systems*, are completed with a stiffening beam. This avoids the use of an excessive number of cables.

Cable-stayed systems are often used in cantilever systems with the stiffening beam in compression. For this reason these are not purely cable systems, but rather hybrid systems with cables and elements in compression. We will discuss cable-stayed systems in the chapter devoted to arch-cables systems.

Cable networks,
tents and membranes

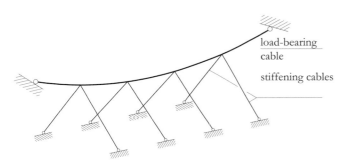

Load-bearing cable and stiffening cables in space

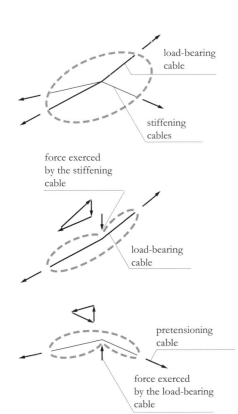

Internal forces in the cables under the effect of the pretensioning

In the examples examined thus far, the load-bearing cable and its stiffening structure were always arranged in a single, usually vertical plane. But it is possible to stiffen a cable by means of a system of cables that intersect the main plane. The result is an extremely rigid system, even when the pretensioning forces are relatively small. The figure represents the functioning in the case of pretensioning of the system. The two internal forces in the load-bearing cable and the force of contact exerted by the stiffening cable act on the subsystem isolated around the load-bearing cable. The three forces are in equilibrium. In other words, the force exerted by the stiffening cable corresponds to the deviation of the internal force in the load-bearing cable. The same consideration can be applied to the subsystem formed around the stiffening cable, based on the principle that the deviation of the internal force in the load-bearing cable must be balanced by the deviation of the internal force in the stiffening cable. A cable stabilized in this way also has another advantage: the system is capable of carrying loads that do not necessarily act in the plane of the load-bearing cable, like wind pressure, for example.

Multiple load-bearing cables are required to support a roof. They can be parallel, as in most of the real examples described thus far, or they can be placed at varying distances. The row of stiffening cables seen previously can intersect all the load-bearing cables in order to stabilize them.

Systems of cables in space

Cable networks

In this way we obtain a cable network in which, exactly as in the case of a cable beam, the load-bearing cables have an upward curvature, while the stiffening cables have an inverse curvature and push downward, as illustrated in the adjacent example. The static functioning is also identical: the stiffening cables exert a downward thrust that, combined with the permanent loads, is taken up by the load-bearing cables. The variable load, if aimed downward, causes an increase of the internal force in the load-bearing cables and a decrease in the stiffening cables. Vice versa, if the variable load pushes upward, due to wind suction, the internal force in the stiffening cables will increase, while the load-bearing cables will slacken.

Cable networks, with the membrane that covers them, form surfaces in space. The spatial figure that corresponds to the parabola in the plane is the hyperbolic paraboloid described by the equation $z = c_1 \cdot (1 + x/c_2) \cdot (1 + y/c_3)$. The name comes from the fact that the horizontal sections of the surface are hyperbolas, while the vertical sections form parabolas. Thanks to this characteristic, the force of deviation of the stiffening cables is uniformly distributed along their length. In the hyperbolic paraboloid, as in all surfaces of this type, it is possible to cut the geometric figure with two vertical planes in well-defined directions, so that the curve degenerates to a straight line (the so-called generating straight line). In the case of the hyperbolic paraboloid determined by the equation indicated above, these sections are characterisized by x or y being constant. If the two sets of cables are arranged along these straight lines they will be without curvature, so it will be impossible to create an interaction between the two sets in order to stiffen them. Furthermore, the load-bearing cables, with their infinite slenderness ratio, are capable of carrying loads only after significant deformations.

Once the form of the surface has been defined, it is then indispensable to specify the directions of the load-bearing cables and the stiffening cables for which the curvatures are never zero. The greatest efficiency is achieved when the upward curvature of the load-bearing cables and the downward curvature of the stiffening cables are at a maximum. This happens when the cables are arranged along the so-called main directions of curvature. The two main directions, where the curvature is maximum and minimum (in this case negative maximum), are perpendicular to each other for any surface.

Stadium of Saint-Ouen, France, 1968-71, Arch. A. Kopp, Eng. R. Sarger

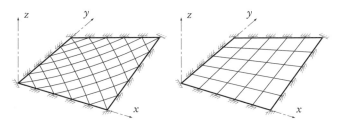

Hyperbolic paraboloid: case with cables arranged along the main directions of curvature and case with cables along the generating straight lines: the first case is much more rigid.

Roof of the Olympic stadium of Munich, Germany, 1967-72, Arch. Behnisch & Partner + Frei Otto, Eng. Leonhardt & Andrä, J. Schlaich

As we have seen, the particular feature of the hyperbolic paraboloid is that it has parabolic sections with uniformly distributed forces of deviation. By varying the distribution of these forces we can obtain a limitless number of other possible surfaces that – as long as they have two main curvatures, one upward, one downward – can function as cable networks. The roof of the Olympic stadium of Munich is a clear example of the freedom of form offered by these structures.

The cable network can be anchored to a rigid structure, as in the case of the stadium of Saint-Ouen, or it can be attached to another funicular system, as in the example of Munich. In the latter case the edge cable, loaded by the internal forces of the load-bearing and stiffening cables, will carry the tensile internal force directly to the supports, composed of foundations and pylons. The illustration shows another case where the cable network is supported by a pylon and anchored to the ground. In the examples described above, the cable network constitutes the main load-bearing structure on which a secondary structure is fastened.

Konzertmuschel at Radolfzell, Germany, 1989, Eng. and Arch. Ingenieurplanung Leichtbau IPL

Tents and membranes

If the cables are brought together and intertwined, the cable network is transformed into a tent composed of a fabric. This structural solution has been known to man since prehistoric times: using animal skins or cloth attached to the ground and supported by wooden poles, man learned how to build very efficient structures.

Apart from structures that required quick, frequent assembly and knock-down, such as circus tents, until the start of the 1960s this type of structure did not raise much interest. With the advent of synthetic materials resistant to large internal forces and extreme weather conditions, especially polyester coated with PVC and Teflon reinforced with fiberglass, membranes began to find more interesting applications.

Traditional cloth tent

The structure shown here, with its 425,000 m² of indoor floorspace, is undoubtedly one of the most spectacular examples. The time required for assembly is particularly impressive: just three months for the whole structure.

All these structures have a very low self weight. The main internal forces are therefore due to pretensioning and variable loads. The structural functioning is exactly like that of the cable network. While in the cable the internal force N was measured in kN or Newton, and the tensile stress in the material s was obtained by dividing N by the area of the cross section, in the membrane we will have an internal forces per unit of length n with the units kN/m or N/mm, while the stress in the material corresponds to the internal force n divided by the thickness of the membrane.

If we analyze the stresses by isolating a subsystem composed of a small rectangle of membrane, we see first of all that the element is subjected to two stresses acting in different directions. This should not surprise us, if we consider the fact that the element ideally takes the place of two segments of load-bearing and stiffening cable of a cable network.

Unlike the cable networks beam, in membranes the stresses do not have predetermined directions. On the other hand, as in a cable network, where it is useful to arrange the load-bearing and stiffening cables following the main directions of curvature, so in the membrane we obtain better performance if we introduce the main stresses in the directions that correspond to those of the maximum and minimum curvatures. These stresses are applied to the membranes by imposing movements of the rigid elements that function as supports, or by tensioning the edge cables to which the membranes are attached.

In the design of a tent, then, exactly as happens in the case of the cable network, we need to choose a form that has at every point two main curvatures oriented in opposite directions.

Airport Jeddah, Terminal Haj, 1981, Saudi Arabia, Arch. Skidmore, Owings & Merrill, Eng. H. Berger (210 elements with a square plan 45 × 45 m)

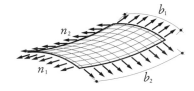

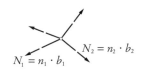

Internal forces in a small rectangular element from a membrane; N_1 and N_2 are the internal forces in an analogous cable network

King Fahd Stadium, Riyadh, Saudi Arabia, 1985, Arch. I. Fraser, Eng.
A. Geiger, H. Berger and J. Schlaich

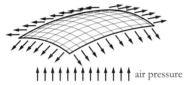

air pressure

Functioning and internal forces on a pneumatic membrane segment

air
pressure

Simplification replacing the membrane element by two cables

Radom, 1946, Eng. W. Bird

This condition regarding the curvatures is very important, not just for the direct influence on the form, but also because it generally requires the use of a whole series of support struts and additional cables to attach and stabilize the desired form. In the example shown here the membrane is stretched by the edge cables and a series of load-bearing cables (known as crest cables) and stiffening cables (known as base cables). These additional cables produce a rippled surface that has the required curvature.

Pneumatic membranes

While in membrane tents the loads are carried by the load-bearing funicular system and the stabilization of the form is ensured by the stiffening system, in the case of pneumatic membranes, the loads are supported by the pressure in the air inside, while the stresses in the membrane have a stiffening function. Again in this case, the membrane is stressed in two directions, but because they have this stiffening function the stresses in both directions exert a downward action. The geometric form will also, necessarily, differ: the two main curvatures, in any case, have to be directed downward. The functioning can easily be understood if we imagine replacing the membrane by a system of cables. In this case, the effect of the internal pressure will be concentrated in an upward force acting on the node where the two cables meet.

The first structures of this type were built immediately after World War II by W. Bird, with the aim of protecting radar installations. Once again, this development was made possible by the introduction of sufficiently strong materials. The absence of facades practically excludes the influence of deformations, so it is possible to use materials that have a low modulus of elasticity. Nylon fibers, in fact, are often used, covered with neoprene rubber.

Because the load-bearing function is performed by the internal pressure, it is indispensable to constantly guarantee this effect. The size of these structures and the presence of accesses makes absolute impermeability practically impossible. It is therefore indispensable to continuously pump air inside. This means that these structures require continuous energy consumption in order to function.

Structures of this type often have a low slenderness ratio ℓ/f. In this case it is indispensable to reinforce the pneumatic membrane with a series of cables. The two sets of cables will both be stiffening and will push the membrane downward, countering the internal air pressure. This makes it possible to cover very large spans. In the Pontiac Silverdome, illustrated here, a membrane of this type covers a stadium that seats 80 000 persons.

Pontiac Silverdome, Detroit, Michigan, 1975, arch. O'Dell Hewlett & Luckenbach, Eng. D. Geiger ($\ell = 216 \times 162$ m, $f = 15.2$ m)

Pavilion US at the Universal Exhibition of Osaka, Japan, 1970, Arch. L. Davis, S. Brody, S. Chermayer, Geismer et Harak, Eng. D. Geiger

Orchids pavilion at Mukôgaoka near Tokyo, Japan, 1987, Arch. Y. Murata, Eng. M. Kawaguchi

High-pressure pneumatic membranes

High-pressure pneumatic membranes, on the other hand, function in a way that is fundamentally different from the system described above. For an immediate, intuitive grasp of their functioning, we can think about the inflatable life preservers and air mattresses used on beaches. In fact, the action is similar to that of the beams we will examine later on.
The Fuji Pavilion, shown here, is a case of a rare structural application of this system.

Pavilion Fuji at the Universal Exhibition of Osaka, Japan, 1970, Arch. Y. Murata, Eng. M. Kawaguchi

Arches

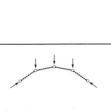

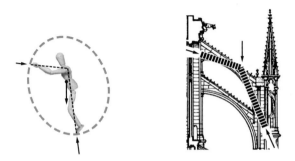

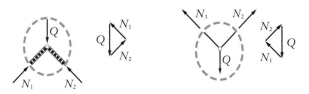

Two examples of structures under compression

Carrying of loads in structures under compression and under tension

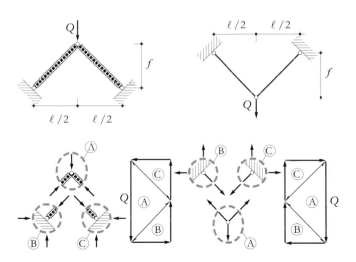

Structure subjected to compression, cable with the same ℓ/f ratio, subsystems and Cremona diagram for a structure under compression and a structure under tension

The next step on our path is the study of structures subjected to compression, previously hinted at with the example of the man leaning against the wall, similar to a rampant arch of a gothic cathedral (see p. 33).

If we analyze how the load is carried in these two specific cases, for example by considering the subsystem of the structure in the vicinity of the applied load, and we compare its functioning to that of structures in tension, we see that there are some strong similarities.

In all structures under tension, the structure carries the load by deflecting until the point in which the force of deviation of the stresses, whose lines of action correspond to the segments of the cable, corresponds to the load itself.

In structures under compression, the load is still in equilibrium with the two stresses that are deviated, but the form the structure must assume differs. In this case the shape of the structure will be concave. If we complete the structure subjected to a single load by introducing supports, we get a configuration that is very similar to that of the cables. Analysis of the stresses and the support reactions can easily be done with the help of the free bodies and the Cremona diagram. The diagram is also quite similar to that of the cables. If the structure under compression has the same slenderness ratio ℓ/f as the cable, so that the slopes of the poles are identical, the length of the vectors represented in the two Cremona diagrams will also be identical. Therefore we can deduce that:

- the vertical forces on the supports are identical to those of the cable;
- the horizontal forces on the supports have the same intensity, but are pointed in the opposite direction;
- the internal forces also have the same intensity, but are in compression rather than in tension.

If we introduce a change of sign, the equations derived for the cable (see p. 40) will therefore still be valid.

Cases with multiple loads or distributed loads: arches

The similarity between cables and structures subjected to compression is valid, independent of the type of load applied. Once again, in structures under compression, the funicular polygon conserves all its importance: the ideal form of an arch will be identical to that of a cable subjected to the same loads, but simply turned upside-down. The illustrations show examples with multiple or distributed loads.

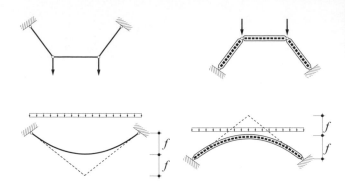

Analogy between cable and arches and case with two symmetrical loads, case with uniform horizontally distributed load

Parabolic arches

When the load is uniformly distributed, the form of the funicular polygon will still be that of a parabola. Again in this case, the formulae derived for the cable (see p. 46 and 47), if the signs of the internal forces are changed, remain valid.

Catenary arches

If the load is constantly distributed along the length of the arch instead of horizontally, the resulting form will be a catenary curve. As we have seen, the catenary curve is very similar to the parabola. They differ, above all, near the springing (the zone of the supports) where, due to the greater slope, the load is also greater when measured horizontally.

In arches the section is often adapted to the stresses. Because they increase as they approach the springings, the intensity of the permanent loads will be greater in these zones. The funicular polygon will therefore undergo an increase in its curvature at the springings, not only with respect to the parabola but also with respect to the catenary curve. In the example in the illustration we can see the variation of the cross section, which has the effect of further increasing the permanent load at the springings. As we will see later on, this variation of thickness is not only a response to the distribution of the axial internal force, but also has the function of improving the behavior of the structure subjected to variable loads.

Gateway Arch, St. Louis, Missouri, 1965, Arch. E. Saarinen, Eng. F.N. Severud (ℓ = 192 m, f = 192 m, ℓ/f = 1.00)

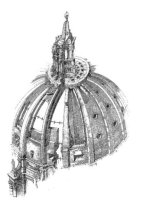

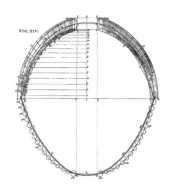

Dome of St. Peter's in Rome, Italy, ca. 1585, Arch. G. Della Porta and D. Fontana , (ℓ = 42 m, f = 26 m)

Analysis by G. Poleni with graphical statics and analogy with the cable (*Memorie istoriche della gran cupola del tempio vaticano*, 1748)

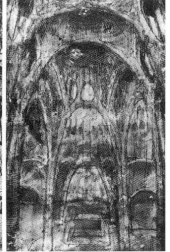

Funicular model of the Colonia Güell chapel in Barcelona, Spain (figure turned upside-down), 1898-1915. Arch. A. Gaudí

Sketch of the support structure composed of arches

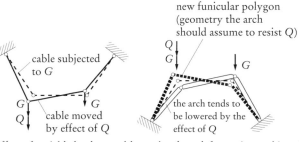

Effect of variable loads on cables and arches: deformation and instability

---

The similarity between structures under tension (cables) and under compression (arches) has often been used to help understand the functioning of arches.

Until the 18th century, many builders looked for a rule to determine the form to be given to arches, for a better response to structural needs. The solutions, for the most part, were empirical: the result was found by means of models, and experience played an important role. The problem was only correctly formulated when the study of funicular systems began, toward the end of the 18th century. The theme became timely due to the interest of mathematicians of the day (Leibniz, Johann and Jacob Bernoulli). The first practical application was made by the mathematician and engineer Giovanni Poleni, who was asked to check the stability of the dome of St. Peter's in Rome, and to explain the cause of the cracks that had formed by analyzing one of its large ribs. To do this, he used a model composed of a cable loaded by spheres, each with a weight proportional to that of the respective quoin. The illustration shows an example with the original construction of the funicular polygon.

The ease with which it is possible to construct models of cables and the fact that they automatically arrange themselves into funicular polygons have often prompted architects and engineers to find the form to give to arches by using the similarity between the two structures, making this approach a valid design aid.

Similarity between cables and arches

When multiple loads that vary in a non-proportional way act on a cable, the result, as we have seen, is a variation of the funicular polygon. If we consider the example shown in the illustration, we see that under the influence of the variable load Q the cable moves closer to the new funicular polygon, to the point of reaching it. So the problem is connected only to the service ability limit state caused by the movements, while a position of equilibrium can always be found.

Influence of variable loads

| **Instability of arches** | If we analyze the analogous case of an arch, one fundamental difference is evident: with the increase in the variable load Q the bars near the load tend to lower, while the new funicular polygon shifts upward, as it is only in this way that the deviation of the internal forces can compensate for the additional load. Without external intervention, for example pushing the arch upward until it reaches the new funicular polygon, equilibrium can no longer be achieved, due to the distancing of the structure from its condition of equilibrium, even through the effect of a minimal additional variable load. In this case, then, we are looking at the phenomenon of *instability*.

The problem is therefore linked to the ultimate limit state: an instability of this type leads to the collapse of the structure. The introduction of constructive provisions to stabilize arches is therefore absolutely indispensable. |

| **Provisions to stabilize arches** | In the case of cables, given the same variable load, increasing the intensity of the permanent load reduces the movement, because the deviation of the funicular polygon depends on the ratio between the two types of load (see p. 53). In fact, in arches this type of provision would also reduce the movement of the funicular polygon; but since even with a very small difference equilibrium cannot be achieved without external intervention, again in this case collapse is inevitable.

For the same reason, the solution with a stiffening element similar to the one applied in the case of cables cannot lead to any increase in the stability of the arch. |

| **Addition of stabilizing bars** | As previously demonstrated (see p. 56), a load-bearing cable can be made more stiff by adding a series of stiffening cables. In arches, too, the insertion of additional bars can be an efficient solution to improve stability. The examples illustrated here show the case of an arch loaded by two concentrated forces, in which case the required stability can be achieved by inserting a supplementary cable or a strut. The supplementary strut is capable of preventing downward movement where the variable load acts. Because the arch tends to rise in the least loaded zone, the addition of a cable that constrains it in that zone and prevents it from rising can also ensure the stability of the structure. |

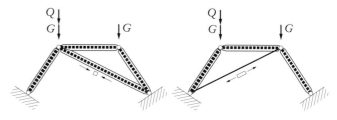

Bars added to stabilize the arch

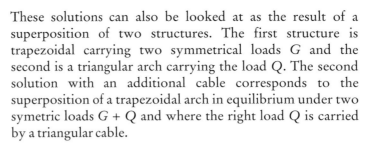

These solutions can also be looked at as the result of a superposition of two structures. The first structure is trapezoidal carrying two symmetrical loads *G* and the second is a triangular arch carrying the load *Q*. The second solution with an additional cable corresponds to the superposition of a trapezoidal arch in equilibrium under two symetric loads *G* + *Q* and where the right load *Q* is carried by a triangular cable.

When an arch is subjected to multiple loads, other bars can be added. Usually stabilizing cables are applied to obtain a light, transparent structure. The examples here show two very slender arches in which the required stability is guaranteed by a series of cables attached to the supports of the arch.
The equilibrium of these structures, easily analyzed by means of free bodies, will be studied later.

Roof of the GUM department store in Moscow, Russia, 1889-93, Eng. V. Suchov

Roof of Chur station, Switzerland, 1988, Arch. R. Obrist + R. Brosi, Eng. P. Rice, (ℓ = 52.10 m, f = 10.00 m, ℓ/f = 5.2)

Val Tschiel Bridge at Donath, Switzerland, 1925, Eng. R. Maillart (ℓ = 43.20 m, f = 5.20 m, ℓ/f = 8.3)

Insertion of a stiffening beam

The solution of adding a stiffening beam, as was shown previously in suspension bridges, can also be used to stabilise an arch. In this case the function is based on two principles: the beam carries non symetrical and non distributed loads and prevents significant movements that would contribute to distance the geometry of the arch from the funicular polygon of loads.
In the arch bridge shown here, thanks to the effect of the deck that functions as a stiffening beam, the reinforced concrete arch has sufficient stability, in spite of the thickness at the key of only 23 cm.

Stiffening the arches by increasing the thickness

A structural solution for the stability problem is to combine the stiffening beam described in the previous solution with the arch, obtaining a structure that is capable of resisting movements. A more efficient solution can be obtained by combining the two functions in the same element. In this case, the arch must have a resistance to compression sufficient to resist the internal force, and a flexural stiffness sufficient to limit the movements. This is achieved by increasing the thickness of the arch and eliminating the hinges. This provision also automatically solves the second problem: if the arch has a sufficient thickness, a load-bearing system can be established within the thickness of the arch that follows the new funicular polygon, so that the arch does not have to move to follow the position of the funicular polygon. Therefore it is not the arch and its form that adapts to the funicular polygon (moving due to the effect of variable loads), but the line of action of the internal forces that must always coincide with the funicular polygon of the loads.

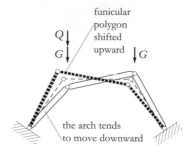

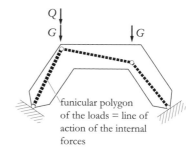

Stiffening of the arch by increasing the thickness: movements are reduced and the funicular polygon can shift inside the material

Line of action of the internal forces

In cables, as well as in the columns covered thus far, the line of action of the internal forces always coincided with the barycentric line of the sections. This was due to the fact that the stresses were of constant intensity in the sections. In the arch just described, the line of action freely shifts inside the material. To better understand this situation, look at an example with a prismatic element subjected to compression. If the line of action of the two forces applied and, therefore, also that of the internal forces passes through the center of gravity of the sections, the stresses in the material will be uniformly distributed throughout the section, as represented in the free body shown here.

If, on the other hand, we move the line of action of the forces and of the internal forces, we will undoubtedly have a different distribution of the stresses inside the material. We will not focus on the exact distribution of the stresses (a problem above all for engineering), but we can try to understand what limits the line of action of the internal forces can reach without causing failure, when we are dealing with a material that only resists to compression (as is the case for masonry).

As we have seen, any material fails when its maximum strength is reached, defined as a maximum stress in tension or in compression. So we can determine the minimum quantity of material required to resist the internal force of compression present inside our prism. The criterion of the ultimate limit state allows us to find the necessary area (see

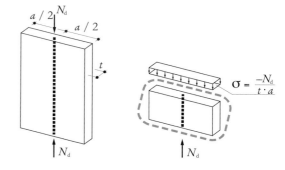

Line of action of the forces corresponding to the barycentric line of the sections

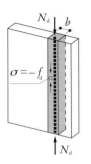
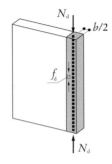

Eccentric line of action of internal forces

Situation of maximum eccentricity

line of action of internal force

Area required to transmit the compressive internal force to the ground and position of the line of action of the internal force

p. 28). This is defined by the equation $A_{req} = N_d/f_d$, where N_d is the internal force augmented by the partial safety factors and f_d is the reduced strength of the material. In our prismatic element we can imagine a zone with thickness of material t and width $b = A_{req}/t$ that resists the internal force N_d, while all the remaining material is not stressed. Clearly this is a simplification that permits us, however, to understand the functioning of structures loaded in this way. Furthermore, when the material has sufficient ductility, this approach is not an approximation, but simply represents an extreme case according to the theory of plasticity. In this way, we can determine the effective strength of the structure.

The line of action of the internal force can therefore approach the surface of the material to the point of reaching a minimum distance equal to half the minimum thickness b of the compressed zone. If the line of action is even closer to the surface, the area under compression will diminish, so there will be an increase of compressive stress that will lead to failure of the material. Clearly, the case in which the line of action is outside the material is also impossible. The situation would be similar to that of the falling man because, without considering the inertial forces, the line of action of the gravitational force is outside the soles of the feet (see p. 12). Nevertheless, based on what we have just seen, even the position of the line of action inside the sole of the foot can be an insufficient condition. In fact, an area of the foot capable of carrying the compression force is indispensable, and therefore the line of action must pass through the center of gravity of this area. Because the internal force, which coincides with the weight of the person, is limited, the area under compression is also very small, so the resultant can be very close to the edge. The situation is different when we walk on soft, cushioning terrain. In this case the strength of the terrain becomes decisive and the area required to transmit the compressive internal force can occupy most of the soles of the feet. This is why when we walk on the soft sand of a beach the danger of falling is much greater than when we walk on a hard surface.

Possible lines of action of internal forces inside an arch

If we return to our arch, we see that the funicular polygon can get dangerously close to the edge, both in relation to the loads $G+Q$ on the left, and in the zone of the load G on the right. We can also see that – given the geometry of the arch and a configuration of the loads G and Q – an infinite number of funicular polygons are possible, situated between the two extreme cases in which the funicular polygon reaches the safety distance $b/2$ in one of the two critical points. In other words, the effective rise f is bordered by two values f_{max} and f_{min}. Note that this indeterminacy has nothing to do with the choice of the rise we made with the cables. In that case the choice was part of the project; now, on the other hand, the indeterminacy we are discussing depends on the thickness of the arch in which funicular polygons can be established, whose rise is slightly larger or slightly smaller than the nominal rise, usually measured in relation to its middle line.

Statically indeterminate and determinate arches

A structure in which infinitely possible lines of action of the internal forces can arise is called *statically indeterminate*. Further on, we will learn how to distinguish these structures from statically determinate structures, in which for a given configuration of loads only one possible funicular polygon exists. Actually, even in a statically indeterminate structure, there will always be only one state of internal forces that can be described by a funicular polygon with a definite rise f. Determination of this funicular polygon, however, requires the conditions of equilibrium we have studied thus far, but also other considerations related to the behavior of the material, its deformation and the boundary conditions. A structure of this type is influenced, for example, by temperature variations, movements of the supports and possible plastic deformations of the material. We will not investigate these effects here, and we will study the functioning of these structures by looking at the two extreme funicular polygons with their minimum and maximum rise.

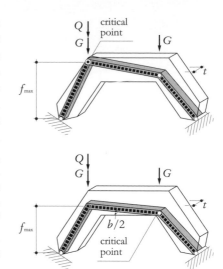

Funicular polygon and zone of the arch under compression: limit with maximum rise and limit with minimum rise

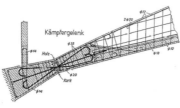

Variation of the funicular polygon in a three-hinged arch due to the effect of a variable load

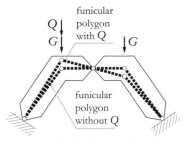

Viaduc de Garabit, Cantal, France, 1884, Eng. M. Koechlin (Eiffel and Co.), (ℓ = 165 m), hinge at the springers

Bridge over the Arve at Vessy Switzerland, 1936, Eng. R. Maillart (ℓ = 55.97 m, f = 4.77 m, ℓ/f = 11.7), hinge at the key and the springers

The fact that the internal forces can depend on the movements of the supports can be considered a severe problem when the arch is set on soft ground. Moreover, especially in the past, when the calculation tools available to engineers were less refined than those used today, the influence of the other factors mentioned could be seen as a major obstacle to the analysis of statically indeterminate arches. For these reasons, engineers have attempted to make statically determinate arches, introducing a so-called *hinge*. This is a cut in the structure with the insertion of a device capable of transmitting the compressive internal force from one part of the arch to the other. This constructive solution, which permits the two parts of the structure to turn freely with respect to each other, introduces a fixed point through which the funicular polygon must always pass.

Note that in the arches we have examined thus far, the supports are also characterized by the presence of hinges. This is why structures of this type that also have a hinge at the key are known as three-hinged arches.

The illustrations show two hinge types. In the example on the left, with a steel arch, the hinge is formed by a true mechanical device, with a stud that permits free rotation of one part of the arch with respect to the other. In the second example, typical of a reinforced concrete bridge, the hinge is created by the weakening caused by two incisions made in the material. The remaining zone, where the compressive internal forces are transmitted, is strengthened with sturdy steel reinforcements.

Optimal form of a three-hinged arch

Arch stability can be achieved by increasing the thickness in the critical zones. In certain zones the funicular polygon can vary greatly due to the effect of variable loads, while in other zones, especially near the three hinges, the variation will be minimal. Therefore it may be reasonable, to save material, to vary the thickness in response to static needs. In this case the design of the statically correct form can be subdivided into two phases. The first phase determines the overall form of the arch based on the funicular polygon of permanent loads, and defines the rise (the span is usually given). The second phase determines the thickness of the arch based on the variations of the funicular polygon caused by variable loads.

In the example illustrated here, we see an arch subjected to a permanent load of constant intensity. The most appropriate general form is therefore that of the parabola, corresponding to the funicular curve of the permanent loads. The effect of a concentrated variable load that can act on any point of the arch will be a series of funicular curves, all passing through the three hinges of the arch. On tracing the envelope of the funicular curves and adding a portion of material sufficient to carry the compressive internal force, we obtain the minimum size of the arch capable of supporting the loads without areas under tension.

The Salginatobel Bridge shown here is a typical example of the form obtained with this procedure. Note the similarity of its form to that of Tower Bridge (see p. 59). In that case the form was obtained by overlaying two cables: the first with the function of carrying loads acting on half the span, the second for the remaining loads. In fact, also in the case of the three-hinged arch, a similar approach can be taken leading to the same results: the different overlaid arches are simply the funicular polygons derived from the variable loads on the two parts of the arch.

In the Austerlitz Bridge over the Seine, shown here, where the arch is composed of a reticular steel structure, the similarity to the Tower Bridge of London is even more evident. In this case, given the fact that the deck of the bridge is lower than the arch, the load is transmitted to the arch not by pillars, as usually happens, but by ties. In any case, the functioning is identical to that of bridges with an upper deck and direct transmission of loads. Another particular feature of this bridge is the position of the two hinges at the springers, which are not located at the supports. The form of the arch, which follows the criterion of maximum structural efficiency, is identical to the one constructed previously in the zone between the two hinges at the springers. In the two segments between the supports and the springer hinges, the thickness has also been varied to permit the shift of the funicular polygon inside the arch, thus leading the maximum thickness at the supports.

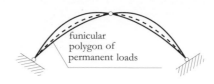

funicular polygon of permanent loads

Optimal form of a three-hinged arch with uniformly distributed permanent load and concentrated variable load

Salginatobel Bridge near Schiers, Switzerland, 1930, Eng. R. Maillart, (ℓ = 90.04 m, f = 12.99 m, ℓ/f = 6.9)

Austerlitz Bridge over the Seine in Paris, France, 1904, Arch. C. Formigé, Eng. L. Biette, (ℓ = 107.50 m)

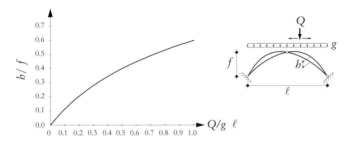

Width h required to carry a concentrated variable load, based on the function of the rise f and the value $Q/g\ell$ (three-hinged arch, without tensile strength)

Alexandre III bridge over the Seine in Paris, France, 1900, Eng. J. Résal, (ℓ = 107.5 m, f = 6.30 m, ℓ/f = 17)

As we have seen, under the hypothesis that the material has to resist only compression, the required thickness of an arch is the result of the eccentricity of the funicular polygon caused by the variable load and the zone of material required to resist the compressive internal force.

If we overlook the latter effect for the moment, and consider only the shift of the funicular polygon, we have a situation very similar to that of the movement of cables caused by variable loads (see p. 53). As in the cable, the necessary thickness depends on two parameters: the slenderness ratio ℓ/f and the *ratio* between variable and permanent loads. In fact, the width of the arch can be expressed in terms of the rise f and the *ratio* between the two loads, as shown in the graph. With a reduction of the rise f, thus increasing the slenderness ℓ/f, the required thickness is diminished, while an increase in the variable load leads to the reverse effect. Remember that what we have just described only applies when the material only resists compression (this is the case of masonry arches or arches in unreinforced concrete). When the structure is composed of reinforced concrete, steel or wood, tensile internal forces can also be involved, and the functioning will differ from what we have described thus far.

The illustration shows a very slender bridge that probably does not fulfill the criterion of the minimum thickness required to prevent the funicular polygon from being outside the section. Later we will examine conditions in which this is possible, as when tensile internal forces can also be carried by the arch. The thickness of the arch, then, does not necessarily have to follow the envelope of the possible funicular curves resulting from the different combinations with variable loads, and therefore it can be significantly reduced.

Nevertheless, we have to consider the fact that the activation of tensile internal forces caused by the funicular polygon being outside the section will be accompanied by a general increase of the internal forces, leading to an increase in the quantity of material required. For this reason, efforts are generally made to limit this situation, by designing the arch in such a way that the general shape corresponds at least approximately to that of the funicular polygon of the permanent loads. The thickness of the arch will also be selected based on two criteria. First of all, a sufficient stiffness is required to keep the structure from being deformed under the influence of variable loads, shifting away from its original position (we will examine this phenomenon, known as stability, later on). Secondly, the thickness of the

Required thickness of a three-hinged arch subjected only to compression

Arches constructed with materials resistant to tension

section must be sufficient to carry the compressive and tensile internal forces that may be generated by variable loads.

Arches whose form does not correspond to that of the funicular polygon of permanent loads

When particular needs demand it, the use of materials that are also resistant to tension, like reinforced concrete, steel and wood, make it possible to choose a form of the arch that is different from that of the funicular polygon of the permanent loads. In this case, we will have a structure known as a *frame*, which is much less efficient than a funicular arch. The example shown here with the intrados of the arch raised to facilitate the passage of the river, demonstrates the possibility of better adaptation to certain functional needs.

Later, we will study the functioning of these structures known as frames. For the moment, let us return to our examination of true arches.

Two-hinged arches, the ideal form

Arches with hinges only at the springers are called *two-hinged arches*. As we have seen, due to their static indeterminacy, an infinite number of funicular polygons can be found inside these structures, with different effective rises. Previously (see p. 78) we described two extremes marked by the compressed zone that touches the edge of the structure at one point: in correspondence to the extrados, when the rise is maximum, and to the intrados, when the rise is minimum.

The thickness of the arch can be optimized by removing superfluous material, so that under a maximum variable load the compressed zone will touch the edge in at least two points. The illustration shows a construction of this type. The overall form of the arch is defined by the permanent loads: in this case, a load distributed with constant intensity generates a parabola. The thickness, on the other hand, is determined by the variation of the funicular polygon caused by a concentrated variable load.

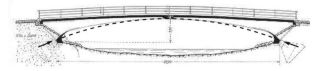

Bridge over the Aar at Innertkirchen, Switzerland, 1934, Eng. R. Maillart ($\ell = 30.00$ m, $f = 3.50$ m, $\ell/f = 8.57$), and funicular polygon of permanent loads

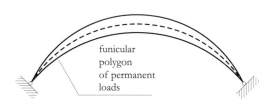

funicular polygon of permanent loads

Optimal shape of a two-hinged arch made with material without tensile strength ($Q = 0.25\ g\ell$)

Maria Pia Bridge over the Douro in Porto, Portugal, 1877, Eng. T. Seyrig, Eiffel and Co (ℓ = 160 m)

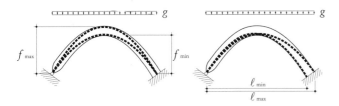

Possible variation of the effective rise and the effective span in an arch with only one hinge (uniformly distributed load; note that only the extremes are represented)

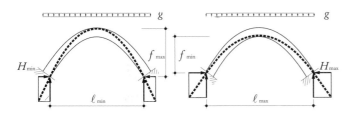

Extreme situations in an arch without hinges, with uniformly distributed load

The Maria Pia Bridge in Porto, shown here, is a classic example of a two-hinged arch. Although the structure is made out of steel, and therefore capable of also carrying tensile internal forces, the arrangement of its thickness approaches the optimal situation just described. Actually, in this case the overall shape of the arch, since it is parabolic, differs from the funicular polygon of the permanent loads. In fact, the loads are transmitted to the arch by means of a few pillars, so the ideal form should be defined by a polygon with its vertices located to correspond with the introduction of the loads. Because this is a railway bridge, the influence of very intense variable loads of great weight with respect to the permanent loads is more important than the effect of approximation of the form. This choice, although not perfectly coherent in terms of static requirements, nevertheless has very convincing visual results.

Arches with one hinge

If the thickness of the structure is sufficient in arches with two hinges at the springers, the line of action of the internal forces can vary, taking on different rises. When, on the other hand, an arch is built with one hinge only, the effective span can vary, as well as the effective rise. Therefore a structure of this type is doubly statically indeterminate.

Arches without hinges

The degree of indeterminacy is even greater in the case of an arch without hinges. In the illustration, for the same arch with a uniformly distributed load, we can see the two limits of the lines of action of the internal forces, which have the following characteristics:
- the first has the maximum effective rise f_{max}, and the minimum effective span ℓ_{min}
- the second has the minimum effective rise f_{min} and the maximum effective span ℓ_{max}

As we have already seen for cables (p. 47), the horizontal component of the thrust corresponds to

$$H = g\ell^2/8f$$

The two values of the horizontal thrust, then, correspond to the two cases just described:

$$H_{min} = g\ell^2_{min}/8f_{max} \text{ and } H_{max} = g\ell^2_{max}/8f_{min}$$

The line of action of the effective internal forces is always located between the two extremes described. It exact position depends, as we have already seen, on many factors that are hard to determine (movement of the supports, temperature, construction method, etc.). A change in one of these factors, such as the movement of the supports, causes a variation of horizontal thrust. If the supports are moved to the point of reaching the maximum thrust, the line of action of the internal forces approaches the extrados at the springers, while at the key it touches the intrados. In an arch without tensile strength, cracks will form on the opposite side, similar to the incisions made in reinforced concrete arches to form the hinges. Further moving of the supports causes further opening of the cracks, which behave like true hinges. In like manner, if we bring the supports closer, we obtain a decrease in horizontal thrust, to the point of reaching the minimum limit.

Reaching one of the two limits we obtain, with the formation of the cracks, the transformation of the system into a three-hinged arch in which the position of the line of action and the intensity of the horizontal thrust are only determined by the loads (statically determinate situation).

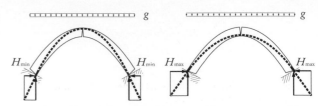

Formation of hinges due to cracking caused by movements of the supports: bringing the supports together, right and distancing the supports, left

Semi-circular masonry arches

In ancient times, masonry arches were often made with a semi-circular form that was significantly different from the funicular polygon of the permanent loads. To make the line of action of the stresses remain inside the arch, the arch had to be built with a sufficient thickness. If we assume uniform distribution of the permanent loads, the thickness h of the arch must be greater than one-sixth of the radius r (more precisely: $h > 0.155 \cdot r$). To this minimum thickness, we must also add the zone necessary to carry the stresses ($b/2 = N_d/2 \cdot f_d \cdot t$). together with a reserve included to compensate for the shift of the funicular polygon under the influence of variable loads.

In reality, the thickness of the arch, even in the most massive historical constructions, does not always reach this limit value. Where the line of action of the internal forces approaches the edge, wide cracks are formed and the parts of the arch rotate with respect to the others, as if there were a hinge at that point. Unlike the case examined previously, here we have the formation, as seen in the illustration, of five hinges, so the system becomes unstable. In fact, the four segments of the arch can freely rotate, distancing themselves from the form of the funicular polygon.

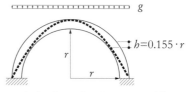

Semi-circular arch, minimum width

Semi-circular arch at Hadrian's Villa, Tivoli, Italy, 125-133 AD

Failure mechanism when the width is insufficient and the line of action gets too close to the edge of the arch

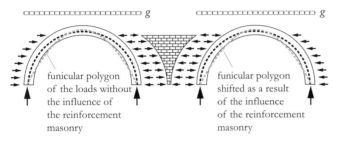

funicular polygon of the loads without the influence of the reinforcement masonry

funicular polygon shifted as a result of the influence of the reinforcement masonry

Activation of the reinforcement masonry by the movement of the arches

If we carefully observe the structures passed down to us by history, we can see that on an arch there are often tympana constructed with reinforcement masonry. This element, due to the effect of the movement of the arches just described, is horizontally compressed and thus exerts a horizontal thrust that modifies the funicular polygon of the loads and brings the line of action of the internal forces back inside the section of the arch.

So the reinforcement masonry plays a fundamental structural role, and its removal can cause the arch to collapse.

The example in the illustration shows a famous Roman construction in which the importance of the reinforcement masonry is quite visible. Solutions of this type have often been used in masonry arches that support bridges, aqueducts or buildings. In the case of porticos, the reinforcement masonry becomes a part of the facade, as can be seen in the case of Palazzo Ducale in Urbino.

Pont du Gard, near Nîmes, France, 19 BC,
(ℓ_{max} = 24.38 m, f = 12.19 m, ℓ/f = 2.0)

Palazzo Ducale in Urbino, Italy, Arch. L. Laurana, 1470 ca.

*Vaults, domes
and shells*

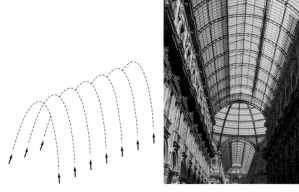

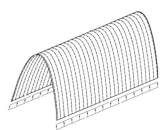

Roof obtained by positioning a series of arches at constant intervals

Galleria Vittorio Emanuele II in Milan, Italy, 1865, Arch. G. Mengoni

Functioning of vaults

Vault covering the audience chamber of the Ctesiphon Palace, near Baghdad, Irak, 3rd-6th C. AD (ℓ = 25.4 m, f = 28.4 m)

Arches as roofing elements

It is possible to roof an area with a series of parallel arches between which a secondary structure is inserted. As in the case of Galleria Vittorio Emanuele in Milan, the secondary structure has the function of receiving the loads distributed over the roof and transmitting them to the arches. If we overlook the possible interaction of the structure with the two terminal facades, and the instability caused by movements perpendicular to the arches, the functioning is identical to that of single arches.

Barrel vaults

If the arches are brought together to the point of touching, making the secondary structure superfluous, the result is a vault (more precisely, a barrel vault).

The distributed load is carried by each of the arches and transmitted, in a uniform way, to the foundation and then to the ground. The foundation or the structure that plays its role must also be continuous, therefore, like the vault.

The problem of instability and absorption of variable loads, whose distribution does not correspond to that of the permanent loads, is identical to that of arches, so the structural solutions described previously can be applied again in this case.

In the example illustrated we see a vault whose thickness is sufficient to guarantee stability and to permit carrying variable loads like wind or seismic tremors that are not too intense (the missing part was destroyed by a violent earthquake). We can see that the general form corresponds well to that of the funicular polygon of the permanent loads. At the springers, where the intrados is practically vertical, the wall is very thick to permit the establishment of a line of action of the internal forces that is inclined due to the effect of the horizontal thrust. This is certainly not the result of quantitative analysis, whose principles were developed more than one millennium after the construction of this vault. Instead, it is the result of profound empirical knowledge and extensive construction experience.

The use of modern materials like reinforced concrete, re_inforced masonry, sheet metal and composite materials permits construction of sufficiently stable vaults by corrugating the surface. In this way, it is possible to limit the thickness to a few centimeters or even millimeters, leading to significant savings on material. The functioning is structurally identical to that of sufficiently thick vaults. In this case the effective thickness, inside which the funicular polygon of the loads can move, does not correspond to the thickness of the material, but to the amplitude of the undulations.

The line of action of the internal forces can, in fact, shift on the corrugated surface, to adapt to the funicular polygon of the loads. The illustration shows two cases where the movement of the line of action, in the channel or at the top of the wave, allows the achievement, respectively, of the minimum rise and the maximum rise.

When the form of the vault does not correspond to that of the funicular polygon of permanent loads, the structural solution used for round arches can be applied, with the activation of reinforcement masonry. In the case of vaults, this function can be performed by simple filler material, which if sufficiently well packed can carry the effect of compression required to bring the line of action of the stresses inside the structure of the vault.

The illustration shows a modern example where a slender masonry vault is stabilized by a covering of concrete.

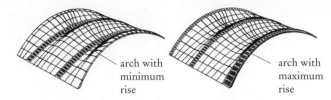

arch with minimum rise   arch with maximum rise

Barrel vault with corrugated surface, functioning with minimum and maximum rise

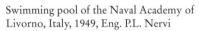

Swimming pool of the Naval Academy of Livorno, Italy, 1949, Eng. P.L. Nervi

Rice silo at Vergana, Uruguay, 1978, Eng. E. Dieste (ℓ = 30 m, f = 15 m)

Blimp hangar at Orly, France, 1923, Eng. E. Freyssinet (ℓ = 86 m)

Storerooms of the Pharaoh Ramses II at Luxor, Egypt, 13th century BC

Villa Sarabhai at Ahmedabad, India, Arch. Le Corbusier

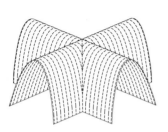

Geometric construction of the groin vault

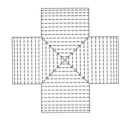

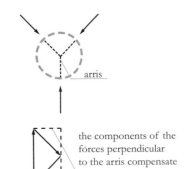

the components of the forces perpendicular to the arris compensate each other

arris

Functioning of a groin vault

Groin vault of the Baths of Diocletian in Rome, Italy, 300 AD ca., transformation into the church of Santa Maria degli Angeli by Michelangelo in 1561

By crossing two perpendicular barrel vaults and removing the portions blocking passage below the point of intersection at the top, a new structural figure is obtained, known as the groin vault (or cross vault).

In the central zone we can see the four segments of barrel vault and the four corners, or arrises, corresponding to the intersections, which also play a structural role.

Where the vaults intersect, the arches cannot transmit their thrust to the ground, but must rest on the corners. As the equilibrium study of this zone shows, for each arch the horizontal component of the perpendicular thrust at the diagonal rib is balanced by that of the other arch that reaches the rib at the same point, while the component in the direction of the rib, like the vertical component of the thrust, is introduced in the rib itself.

Note that all the elements discharge their thrust to the ground at just four points. This means that a groin vault, unlike a barrel vault (which must be supported by a continuous structure) can be placed on single-point supports.

The first major applications of this structural form were made by the Romans. In the Basilica of Massentius the groin vault, built with concrete in 30 AD, has a large span of 25 m. The roof of the baths in Hadrian's Villa at Tivoli, built in about 120 AD, also has a groin vault.

Groin vaults are usually much more stable than barrel vaults. This is because the two vault elements can cooperate to stabilize each other. Each element has a function similar to that of the reinforcement masonry placed on an arch.

Groin vaults developed significantly during the Gothic period. Alongside the characteristic pointed arch form, tripartite groin vaults were introduced. Their geometry, with six vault segments, is the result of the intersection of three barrel vaults, whose crests meet at one point.

The roofing of the central nave of the cathedral of Beauvais, shown here, is a very good example of this type of geometric construction.

The apse is covered by a vault whose form follows the same principle. Here as many as eight vaults intersect. The result is a structure supported by a series of pillars placed on a circumference.

These pillars, due to their extreme slenderness, are capable of carrying only the vertical component of the thrusts, while the horizontal component must be countered by a series of rampant arches in a radial arrangement (clearly visible in the photograph of the exterior).

Cathedral of Beauvais, France, 1337 (roof reconstructed after the collapse in 1284 caused by insufficient stability)

Fan vaults

A further evolution of the groin vault happened in England, in the 14th century. The vault elements and the arrises were combined in a continuous structure that was rounded, even in its plan. The geometric parts of this new structure are called conoids, and they have a generatrix curve different from that of the common cone, obtained through rotation of a generating straight line.

The contrast between the groin vault and the new form is clearly visible in the illustration. A fan vault is usually formed by multiple conoids placed side by side. They intersect at the longitudinal and crosswise crests, where there will be a rib or a horizontal element, known as a *spandrel*.

Project for the Cathedral of New Norcia, Perth, Australia, 1958, Arch. P.L. Nervi, A. Nervi, F. Vacchini, C. Vannoni ($\ell = 35.40, f = 30.85$ m)

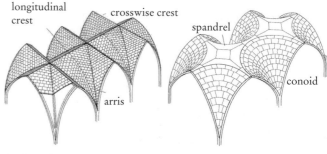

longitudinal crest   crosswise crest   spandrel   conoid   arris

Groin vault   Fan vault

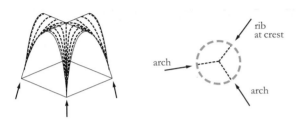

Functioning of the fan vault

Chapel of King's College at the University of Cambridge, UK, 1515
(ℓ = 12.7 m)

Centre des nouvelles industries et technologies in Paris, France, 1958,
Arch. R. Camelot, J. De Mailly, B. Zehrfuss, Eng. N. Esquillan
(ℓ = 218 m along the fassade, f = 46.30 m)

Considering the fact that a conoid is a surface with a double curvature, in structural terms we should thus be talking about a shell, instead of a vault; nevertheless, these structures are usually called *fan vaults*. In terms of static functioning, in any case, the definition is apt, because arches are effectively formed on the surfaces that transmit internal forces in one direction only, as in the case of vaults. These arches are arranged radially, like the sticks of a fan.

If the geometry of the arches does not correspond to that of the funicular polygon of the loads, compression or tension zones will also have to form in a horizontal direction, as we will see later on in the case of domes. In this case the functioning comes closer to that of shells.

Where the conoids intersect, the arches, with which we can describe the functioning, meet and transmit a part of the internal force to the rib at the crest, or the spandrel. One modern example of a fan vault is the roof of the Centre des nouvelles industries et technologies in Paris. This is a reinforced structure concrete of impressive size, with a triangular plan. To increase stability and to limit the influence of variable loads, the structure is composed of two corrugated reinforced concrete shells, with a thickness of just 6-12 cm, placed at intervals varying from 1.90 to 2.75 m.

In the view from above it is easy to see the ribs located on the crests, which have the function of carrying part of the internal forces in the fan arches.

Pavilion vaults

If we consider only the part of the barrel vault we have removed in the geometric construction of the groin vault, we obtain another structural form, known as the *pavilion vault*. Again in this case, we have four barrel vault parts, known as spindles, and four arrises, corresponding to the intersections. If we analyze the functioning, by imagining arches arranged on the four spindles, we can see that they can directly discharge their loads to the ground. This means that just as in the case of the barrel vault, but unlike that of the groin vault, pavilion vaults must be placed on supports that are continuous along the whole perimeter.

In the upper part the arches are interrupted by the arrises. Their trust is partly countered by that of the arch that reaches them from the other spindle, and partly introduced in the ribbing. Unlike a groin vault where the ribs were pushed downward by the thrusts of the arches, in pavilion vaults they are pushed upward. Their internal force is zero at the point of the supports and increases, heading upward, little by little as they receive the thrust from the arches, until reaching the maximum compressive internal force at the key.

Pavilion vaults can be combined with barrel vaults, to cover very long rectangular spaces. In this case the compressive internal force in the arrises, which as we have seen reaches its maximum at the key, must be countered by a compressed element located on the crest of the barrel vault.

As in groin vaults, pavilions vaults can also be made by the intersection of any number of barrel vaults. The result is a pavilion vault whose perimeter is a polygon with any number of sides. The illustration shows for example a pavilion vault with an octagonal plan.

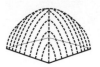

Geometric construction of the pavilion vault

 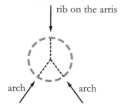

rib on the arris

arch   arch

Functioning of the pavilion vault

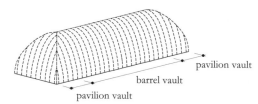

pavilion vault

barrel vault

pavilion vault

Pavilion vault made with a barrel vault

Pavilion vault with octagonal plan

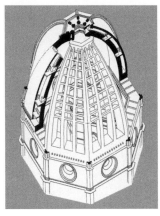

Dome of Santa Maria del Fiore in Florence, Italy, 1436, F. Brunelleschi (ℓ = 42 m)

	Structure supported at a few points	Structure supported along the entire perimeter
Surfaces with single curvature, bordered by arrises: functioning similar to barrel vaults with arches arranged in parallel	Groin vault	Pavilion vault
Continuous surface with double curvature, with radial arrangement of arches	Fan vault	Dome

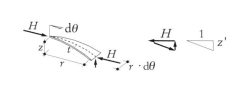

Dome functioning simply as a set of arches, internal forces on the arches

The dome of Santa Maria del Fiore in Florence is actually an example of a pavilion vault with an octagonal plan. The support structure is composed of two masonry vaults: one external vault, with a thickness of just 0.58 m, and an internal vault whose thickness varies from 2.40 m at the base to 2.10 m at the key. The two vaults are connected by eight ribs at the arrises, sixteen intermediate vaulting ribs and horizontal connection elements. As a whole, the structure is about 4 m thick. As seen in the diagram, the intermediate ribbing has a radial arrangement, and does not rest on the ribs placed at the arrises. The latter can therefore discharge part of their thrust directly onto the supports. Although the form is that of a pavilion vault, the presence of radial ribbing makes its functioning close to that of a dome.

Domes

From a structural viewpoint, the term "dome" should be set aside for roofs with a circular or an elliptical plan. We can consider structures of this type as pavilion vaults composed of infinite spindles separated by infinite arrises. In this case, the width of the spindles becomes infinitesimal, so the support elements must be composed of an infinite series of arches, all arranged radially. To better visualize the functioning, we can represent a finite number of arches.

The differences and similarities between domes and other types of vaults are summed up in the adjacent table.

Effective functioning of domes

In the dome model just described, each of the arches will carry the load acting on the corresponding surface. The intensity of the internal force in the arch can immediately be calculated, knowing the horizontal component of the thrust H and the inclination of the dome z'. If we want to know the effective compressive stress on the material (internal force per unit of surface), we have to divide the compressive internal force by the area of the section

$$A = t \cdot r \cdot \mathrm{d}\theta$$

where t is the thickness of the dome, $\mathrm{d}\theta$ is the angle of aperture of the sector of the dome contained in the arch, and r is the distance between the axis of symmetry and the point of the dome we want to analyze. Note that at the level the axis of symmetry, i.e. at the peak of the dome where the radius r is zero, the area of the section of the arch is also zero, so the stress on the material is infinite. A situation of this type is clearly impossible, because no material could resist such a stress.

In fact, in the zone of the peak the loads are not carried only by the arches set radially along the meridians. Compression rings form along the parallels of the dome, capable of discharging the arches and limiting the stresses.

The illustration shows, as a free body diagram, a segment of a dome from the zone of its peak. It is subjected to internal forces both radially (meridian arches) and in the direction of the rings (parallels). As the equilibrium shows, in the force polygon, the internal force of the arch can be entirely introduced in the system of rings.

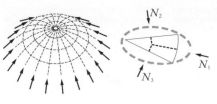
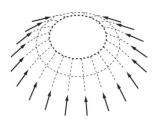

internal forces in the arch in equilibrium with the deviation of internal forces into the rings

Dome functioning as a set of arches and rings

Functioning of a dome with an opening at the key for the skylight

Domes with a central skylight opening

We can even imagine the arches completely shifting their internal forces to the rings. In fact, it is possible to build a dome with a large open portion at the key, without compromising its functioning. In this case the arches are completely interrupted at the peak of the dome. They simply rest on the compression rings located at the edge of the skylight. Clearly these rings will have to carry larger internal forces than those of a continuous dome.

The Romans gained great mastery in the construction of domes, and made widespread use of this possibility. The illustration shows the dome of the Pantheon in Rome, with the characteristic opening at the key, whose diameter reaches 9 m. The arches and the rings visible on the intrados, although arranged in the direction of static action, are purely decorative: the structure is, in fact, composed of a massive concrete dome.

The dome of the Pantheon in Rome, Italy, with the opeion at the key, 128 AD ($\ell = 43.30$ m, $f = 21.65$ m)

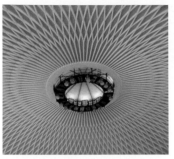
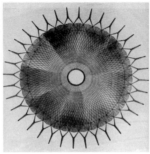

Palazzetto dello Sport, Rome, Italy, 1957, Arch. + Eng. P.L. Nervi, with Arch. A. Vitellozzi ($\ell = 60$ m, $f = 21$ m)

The Palazzetto dello Sport, also in Rome, is a modern application of the same principle. The dome is composed of a very slender shell of reinforced concrete. Again in this case, the ribbing of the intrados should not deceive us: it edges delimites the prefabricated elements and contribute to the stability of the vault. Moreover, in this case the direction of the ribs does not correspond to that of the compression effectively present in the dome (arch-meridians and ring-parallels).

At the edge of the central opening we can see the ring that has the job of carrying the horizontal thrust of the interrupted arches. At its perimeter the dome is supported on 36 Y-shaped trestles, arranged radially and inclined according to the tangent to the dome. In this way, they directly transmit the thrust of the dome to the foundations.

Steel domes

Thus far, we have described the functioning of masonry or concrete domes by means of virtual arches and rings that we can trace inside the material to show how compressive internal forces are carried. In the case of steel domes, given the fact that they do not have a continuous structure, true arches and rings have to exist. In fact the cladding, often in glass, does not generally have a primary support function.

Crossed arches

In the case of steel domes, we can imagine a structure composed only of crossed arches. In fact, for constructive reasons, in the zone where they cross there has to be a connecting element that will transfer the internal forces in all directions.

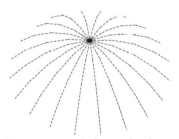

Dome composed of crossed arches

Dome of the Reichstag in Berlin, Germany, 1999, Arch. N. Foster, Eng. Studio Leonhardt + Andrä

Domes composed of arches and rings

Actually, in steel domes it is also useful to insert rings. In this way, the arches are partially stabilized and greater freedom is permitted in the choice of the form, as we will see later on. Another advantage is the possibility of making an opening for a skylight in the central part.

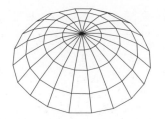 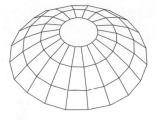

Steel dome with arches and rings   Steel dome with central skylight

Galleria Vittorio Emanuele II in Milan, Italy, 1865, Arch. G. Mengoni

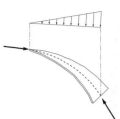

Triangular load on a dome segment and funicular polygon of a dome without compressed rings

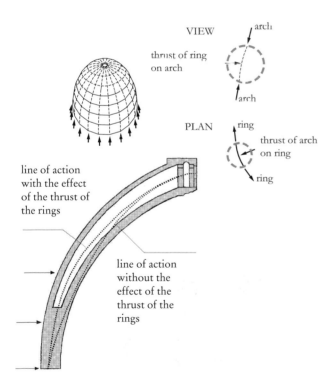

VIEW

thrust of ring on arch

arch

arch

PLAN

ring

thrust of arch on ring

ring

line of action with the effect of the thrust of the rings

line of action without the effect of the thrust of the rings

Functioning of a spherical dome with the action of the rings under tension, funicular polygon of the loads in the dome of St. Peter's in Rome, with and without the action of the reinforcement rings.

If we assume a uniformly distributed load on the horizontal surface and neglect, for the moment, the action of the rings under compression along the parallels, the funicular polygon will take on the form of a third-degree parabola. This comes from the fact that the arch we can cut out of the dome is actually a triangular segment when seen flat. The load increases in a linear way, then, as we move away from the axis.

In the case of a masonry or concrete dome, as we have just seen, the formation only of radial arches is not possible, because the specific stresses on the material would be greater than its strength. So there must necessarily also be compression rings, which also modify the form of the funicular polygon. It can be demonstrated that in this case, assuming a constant intensity of the stresses, in both the direction of the meridians and of the parallels, the funicular polygon will assume the form of a second-degree parabola.

As already described for arches and cables (see p. 73), the determination of the funicular polygon and the design of the ideal form for the arches and domes were only developed at the start of the 18th century. Previously, the form was the result of experience and an empirical approach: arches that turned out to be unstable were reinforced by masonry or a system of rampant arches.

In antiquity, domes were often spherical or close to that form, so the funicular polygon of the loads was dangerously close to the intrados, and the structure had a tendency to shift outward, just as happened with round arches. For this reason, in both the masonry domes of the Renaissance and the concrete domes of the Romans, large radial cracks formed in the lower part of the structures.

One interesting example is that of the dome of St. Peter's in Rome, where studies conducted to find the cause of these cracks led, for the first time, to the construction of the funicular polygon of a dome. The studies showed the fragility of the equilibrium and led to reinforcement of the structure. This consisted of the installation of rings in tension to exert a horizontal thrust on the arches and thus bring the line of action of the internal forces back inside the structure, exactly as happened for round arches, by means of reinforcement masonry. Note that in these domes the inward thrust of the rings can completely compensate the outward thrust of the arches, so that a vertical reaction can be sufficient at the springer.

The example of the dome of St. Peter's demonstrates how a very simple constructive provision makes it possible to achieve optimal functioning, even when the form of the dome does not correspond to that of the funicular polygon. This means that domes can be designed with greater freedom than arches. If the curvature in the direction of the meridians is greater than that of the funicular polygon of the loads, rings under tension are needed, and the material must be capable of resisting the corresponding stresses. Otherwise reinforcement with steel rings will be necessary.

Most of the domes of the Orthodox, Islamic and Hindu worlds have this form. Often the form of the intrados does not correspond to the form visible from the outside, so the difference with respect to the funicular polygon is not as pronounced as it might seem at first glance. In any case, their functioning is possible only through the activation of rings under strong tension in the lower part.

Central dome of the Taj Mahal in Agra, India, 17th century

Conical domes

When the curvature of the meridian arches is less than that of the funicular polygon of the loads, compression rings must also be able to form in the lower part of the dome. One particular case is that of the dome with a conical form, often used for the roofs of towers or steeples. In a cone the curvature of the meridian arches is zero, so there can be no deviation of the compressive internal force in this direction. The functioning becomes immediately understandable if we study a segment cut out of the cone, on which the following loads and stresses are at work: the load with a vertical line of action, an compressive internal force on the base of the segment with the line of action in the direction of the rectilinear meridians, and two compressive internal forces in the direction of the parallel rings. Because the two internal forces acting along the parallel rings refer to two distinct sections, their directions are not identical. Although the intensity is the same, the result will be a form of deviation that pushes the element outward. The equilibrium of these forces can be studied with a force polygon in space, or by first analyzing the horizontal components of the forces and then the projection of the forces on a vertical plane that passes through the axis of symmetry of the cone. We can understand the situation even better by using an analogy, thinking for example about a person leaning against the two walls at the corner of a room.

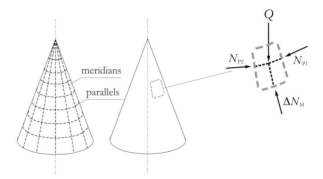

Internal forces in a cone caused by permanent loads

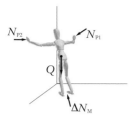

Analogy: a person leaning on the two walls at the corner of a room

Steeple of Sant'Andrea in Mantua, Italy, 1413

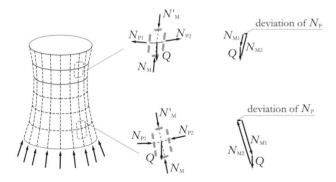

Functioning of hyperboloids of revolution under permanent load

In reality, in the subsystem extrapolated from the dome we should also have considered the compressive internal forces transmitted by the zone above (which would not, however, influence the equilibrium). Instead, the internal force considered represents only the increase that happens in this zone. The internal force along the meridians, then, gradually increases as we move downward.

The formation of an opening in the upper part of the dome, resulting in a truncated cone, would have the sole effect of reducing the compressive internal force along the meridians. The structural functioning described above would however remain the same.

Hyperboloids of revolution

Hyperboloids of revolution, often used for the cooling towers of thermal or nuclear power plants, have similar characteristics. In the lower part the curvature of the arches that act along the meridians is even aimed upward, so the force of deviation of the internal force along the meridians is added to the permanent loads that act in a downward direction. The load-carrying mechanism by means of parallel rings in compression, and the transmission to the supports through the arches along the meridians, is therefore similar to those of the cone.

In the upper part the inclination is inverted, so it acts as an overturned cone with its point at the bottom. The increase of compression along the meridians has a component that acts outward, so the parallel rings are under tension. Thinking back on our analogy, it would be as if the person were now gripping two cables, as represented in the illustration.

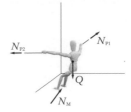

Analogy to explain the functioning in the upper part of the hyperboloid of revolution

The Chandigarh Palace of Assembly shown here is probably one of the most well known examples of architectural use of parabolic hyperboloid reinforced concrete shells.
To facilitate construction of the model rectilinear bars were used, arranged on the generatrices of the hyperboloid of revolution. The real construction is composed of a shell of reinforced concrete

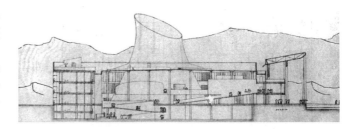

Parliament building at Chandigarh, India, 1952-1962, Arch. Le Corbusier, section and model.

This type of structure can thus be easily realized by using straight bars and circular rings as in the example of the tower by Russian engineer Vladimir Suchov shown here.

A very successful modern example is without doubt the Sendai mediatheque, where a series of steel towers carry the floors and stabilize the whole structure.

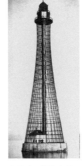

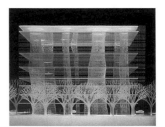

Adzigol lighthouse near Kherson, Ukraine, 1911, Eng. Vladimir Suchov

Maquette of the Sendai Mediatheque, Japan, 2001, Arch. Toyo Ito, Eng. Mutsuro Sasaki

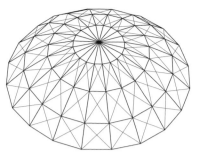

Steel dome with diagonals

"La Goccia" conference room at the Lingotto in Turin, Italy, 1996, Arch. R. Piano, Eng. Studio Arup & Partners

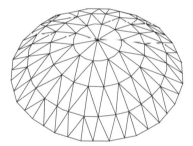

Dome with secondary arches

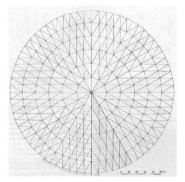

Stadium in New Orleans, Louisiana, (dome with diameter of 207.3 m), 1973, Sverdrup & Parcel and Associates

In tall structures, like the hyperboloids of revolution used for cooling towers or the cones that cover steeples, the controlling load is not the permanent load but the variable load caused by wind thrust. In this case the analysis of the functioning becomes more complex, because the load is no longer symmetrical. There are internal forces that do not follow the meridians and parallels of the structure. In the case of masonry or concrete domes the functioning can still be described using a series of arches and ties arranged on the surface of the structure.

In the case of steel domes, composed of meridians arches and parallel rings, the transmission of the horizontal thrusts and non-symmetrical actions becomes more complex.

One possible structural solution is to increase the width of the arches, allowing the funicular polygon of loads to shift without creating excessively large deformations or tensile internal forces. In the dome of Galleria Vittorio Emanuele in Milan (photograph p. 98) it is easy to see that the arches are capable of resisting non-symmetrical loads.

A second possibility is that of inserting diagonals inside the quadrilaterals formed by the meridians and parallels, as shown in the upper illustration.

These diagonals can be composed of simple cables. The result is a more transparent structure, in spite of the added elements. In this way the section of the meridians arches and parallel rings can be significantly reduced.

Another solution is to use secondary arches capable of carrying the non-symmetrical thrusts. In this case, however, the result is a structure that is relatively opaque, due to the great number of bars needed to resist the compression. Thanks to its efficiency, this type of structure is usually applied in large domes, or when transparency is not an objective.

Carrying of non-symmetrical horizontal or vertical loads

Geodesic domes

In the solutions described above, in spite of the introduction of supplementary bars in the form of diagonals or arches, the meridian arches and parallel rings are still recognizable. In the case of geodesic domes, developed by Buckminster Fuller, three systems of very irregular arches intersect. The form of these domes is based on the projection of the icosahedron on the surface of the sphere. This makes it possible to create spherical domes, as shown in the illustrations.

Grid domes

A structure composed of just two systems of arches was recently developed by the German engineer J. Schlaich. As we will see later, the same principle can be used to obtain structures with various forms. In the dome shown in the illustration the two systems of arches that form the grid each carry a part of the load, and both have a downward curvature. It functions, therefore, in a different way from a cable net structure under tension, where one family with upward curvature carries the loads, while another exerts a downward pretensioning thrust. A grid dome with a square grid, like the ones shown here, does not, however, have enough stiffness to resist asymmetrical loads such as wind or snow. To avoid this problem, slender diagonal cables are added.

Icosahedron

Geodesic dome

Dome of the US pavilion at Expo '67 in Montreal, Canada, Buckminster Fuller

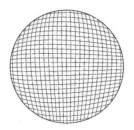

Grid dome

Roof of the swimming pool at Neckarsulm, Germany, 1989, Arch. K.-U. Bechler, Eng. J. Schlaich

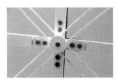

Assembly of the structure and close-up

Test roof at Jena, Germany, 1932, Eng. F. Dischinger and U. Finsterwalder ($\ell = 7.3 \times 7.3$ m, thickness $= 1.5$ cm!)

Gartencenter Wyss in Solothurn, Switzerland, 1961, Eng. H. Isler

The domes discussed so far are characterized by a geometric form that can be obtained by rotating any curve around an axis (usually vertical). These cases are indicated with the term "domes of revolution". In fact, the family of domes is much larger and includes all the geometric figures that do not necessarily conform to this rule: so we can talk about "arbitrary domes" or, to put it more simply, shells. The latter term indicates a spatial structure essentially subjected to compression, whose thickness is limited with respect to the other dimensions. In other words, it is a structure similar to a membrane, but under compression.

Shells and arbitrary domes

The four hyperbolic paraboloids of the structure just described can be assembled in such a way as to be supported at the four corners. The resulting configuration is similar to the fan vault described previously (see p. 93). In this case, only two compressed ribs converge on each support, and the forces they transmit to the supports will also have horizontal components. It is possible to counter these thrusts by means of ties placed between adjacent supports.

As in fan vaults, in the case of hyperbolic paraboloids it is possible to arrange many elements side by side. The result is a roof with many supports, without any size limitations.

The hyperbolic paraboloids just described are all delimited by generatrices, so the edges are rectilinear. In the example shown in the illustration, the edges are instead defined by sections that do not correspond to the generatrices. This makes it possible to have curved edges (parabolas and hyperbolas).

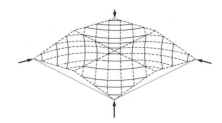

Functioning of a roof composed of four hyperbolic paraboloids supported at the four corners

Capilla abierta, Lomas de Cornavaca, Palmira, Mexico, 1958, Arch. + Eng. F. Candela, wooden centering for the pouring of the reinforced concrete shell (note the similarity with the model by Le Corbusier for Chandigarh), and the completed structure

Monkey saddle surfaces

Like groin vaults derived from the intersection of barrel vaults, *monkey saddle surfaces* are obtained by intersecting hyperbolic paraboloids. It is possible to keep the corners at the intersections, or to smooth them thus creating a continuous surface.

Monkey saddle surface, Los Manantiales restaurant, Xochimilco, Mexico, 1957, Arch. + Eng. F. Candela

Composed arbitrary shells

It is also possible to compose several free-shape shell elements to obtain roofs on several support like in the case of the SICLI plant in Geneva by H. Isler.

SICLI plant in Geneva, Switzerland, 1969, Eng. Heinz Isler

In the case of the Kakamigahava crematorium designed by T. Ito and M. Sasaki, the reinforced concrete shell transforms itself and becomes a column with mushroom on the supports.

Kakamigahara crematorium, Japan, 2007, Arch. Toyo Ito, Eng. Mutsuro Sasaki

Kimbell Art Museum, Fort Worth, Texas, 1972, Arch. L.I. Kahn, Eng. A. Komendant

Cylindrical shells

As we have seen in the case of the hyperbolic paraboloid, the same geometric figure can be used in various ways to obtain very different structures. Their functioning, however, will still have many similarities.

In other cases, structures that appear to have the same form may nevertheless function in a completely different way, if the static conditions at the edges are modified. This is the case, for example, of cylindrical shells, which have the same form as a barrel vault but which unlike a barrel vault are supported at just a few points.

Arches are established in the upper part of the shell, just as in vaults. However these arches, due to the free edges, must transmit their thrust to another carrying system, which acts lengthwise and transmits the load to the supports. This carrying system is composed of longitudinal arches that collaborate with a system of ties, similar to what happens in beams. For this reason, we will return to our study of cylindrical shells later on.

Gridshell structures

We have limited our use of the term shell to continuous structures that in building construction are generally made with concrete or masonry. In other situations, shells can be made of steel, plywood or synthetic materials. Two fields of application are boats and vehicle bodywork. In building, steel or wooden structures of this type are generally composed of overlaid linear parts. In this case, we can use the term gridshell. Great stiffness and good stability can be achieved by crossing three systems of arches. Nevertheless, when maximum transparency of the structure is required, it is preferable to leave out the third system of arches, replacing it with two additional bands of slender cables.

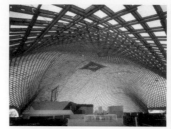

Gridshell structure at Mannheim, Germany, 1971, Arch. Mutschler, Langner and Frei Otto, Eng. T. Happold

Courtyard of the Museum for Hamburg History, Germany, 1989, Arch. Gerkan, Marg & Partner, Eng. Schlaich Bergermann & Partner

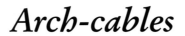

Arch-cables

Opposing structure in the cathedral of Beauvais. On the central nave only the internal roofing is represented, composed of a tripartite masonry cross-vault; the external roof is not visible, and is composed of a wooden structure that only transmits vertical forces to the pillars.

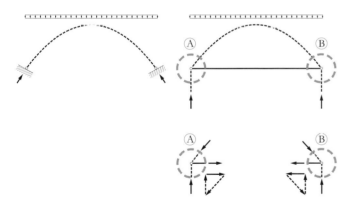

Functioning of an arch with a tie at the springers, as compared to an arch without a tie

As we have seen for arches, vaults, domes and shells, the main load-bearing system is composed of a compression zone that is deviated by the loads and therefore undergoes a variation of slope. So not only a vertical thrust, but also a horizontal thrust acts on the supports, derived directly from the loads and corresponding to the compressive internal force at the key.

If these thrusts cannot be directly transmitted to the ground, for example because the structure is positioned at a certain height, support structures are required. We have already seen how in Gothic cathedrals the vertical component of the thrust is transmitted to the ground by the pillars, while the horizontal component is carried by opposing structures, formed by a complex system of rampant arches and buttresses. These structures also have other functions: they resist the horizontal load of the wind and, in case of earthquakes, they stabilize the entire construction.

But let us return to the main purpose, that of carrying the horizontal component of the thrust. We can imagine a very simple alternative system to perform the same task: a rod that connects the two springers. This solution has often been used, not only to reinforce structures where the opposing structures are insufficient, but also as an initial design choice. The functioning is easy to understand if we analyze the free bodies that include the zone of the springers with the relative supports. The vertical component of the thrust is in equilibrium with the vertical force the pillars exert upward, while the horizontal component is balanced by the internal force the new rod exerts inward. From this analysis, we can immediately see that the rod is under tension: therefore it is a tie. In other words, the force exerted by the tie toward the inside of the arch or the vault is capable of deviating the thrust and bringing its line of action to be vertical at the supports, exactly as happened in the spherical domes, thanks to the rings under tension.

If we analyze a free body that includes the other springer, and if the loads acting on the arch are all vertical, we obtain the same tensile internal force in the tie: this means that it transmits the horizontal thrusts from one part of the arch to the other, balancing them.

Arches with ties

By composing an arch with a tie we obtain a new structure that makes it possible to transfer loads in a different way to the foundations.

The example shows a structure that exploits this principle. It is a temporary centering used for the construction of a reinforced concrete arch. During movement from one span of the bridge to the other, the structure was placed on two barges, clearly capable of carrying only vertical support forces (Archimedes' thrust). The arch of the centering, with just two hinges at the springers, was composed of a wooden structure, while the tie was in steel.

Floating centering of the bridge of Plougastel, France, Eng. E. Freyssinet (ℓ = 170 m)

Fixed and sliding supports

When a structure of this type is placed on the ground it is useful to have supports that permit unimpeded horizontal movement. If this is not possible, the tie cannot lengthen and therefore cannot be activated by the effect of the loads. In this case the functioning is identical to that of a simple arch and the horizontal thrust component is directly transmitted to the ground.

If, on the other hand, both supports are sliding, the system becomes unstable. In fact, a horizontal load acting on the system could not be carried, and would cause the entire structure to move.

For these reasons, it is better to use one fixed support, capable of carrying the horizontal loads on one side, and a sliding support that permits movement and activation of the tie on the other side.

The hinges at the springers of the arches (see p. 79) are a typical example of a fixed support. In these points rotation is possible, while movement – both horizontal and vertical – is prevented. The forces thus transmitted can have any direction, with a horizontal and a vertical component. From this point on we will illustrate these supports in a schematic way, with a small triangle.

Sliding supports are constructions that permit movement in one direction, while they can transmit forces perpendicular to the plane of the sliding. In the diagrams they are shown with a small triangle placed on a circle, representing a roller. In fact, in the past these supports were composed of steel rollers that could roll between two plates: a fixed lower plate, and a mobile upper plate, attached to the structure. Another constructive solution, now also increasingly rare, consisted of a rocker attached at the two ends by joints.

Diagram of an arch with tie, fixed support to the left, sliding support to the right

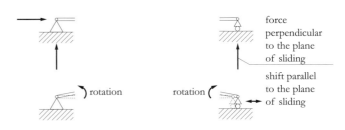

Transmitted forces and possible movements in a fixed support and a sliding support

Examples of sliding supports: roller supports and supports with low-friction materials

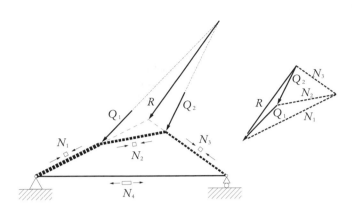

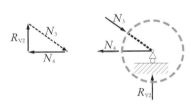

Determination of the funicular polygon and analysis of the arch with tie

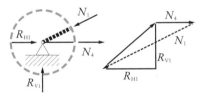

Determination of the force on the fixed support (N_1 and N_4 determined previously)

Modern sliding supports take advantage of the very low angle of friction between certain synthetic materials (Teflon, for example) and metallic surfaces that have been properly polished. With these techniques, the transverse forces do not exceed 1/20 of the force transmitted perpendicularly to the plane of sliding, so we can talk about sliding supports with practically free movement.

If we consider the fact that the main function of a tie is to replace the supports by carrying the horizontal thrust component, the design and analysis of an arch combined with a tie are no different than those of a simple arch. Therefore all we have previously outlined for arches regarding functioning, form, the problem of stability and its possible structural solutions remains valid here.

The geometric construction of the form (funicular polygon) and the analysis can be made by the same procedure: finding the resultant of the loads and its line of action (by means of an auxiliary arch, when the loads are parallel), choice of a funicular polygon of the resultant (corresponding to a triangle) and construction of the intermediate zone by means of the Cremona diagram, from which we can also directly deduce the intensities of the stresses. Clearly this is only true when at least three hinges are present in the arch. When their number is smaller and the arch becomes statically indeterminate, the considerations already outlined for arches apply.

The next step is to find the forces at the supports and the internal force in the tie. This happens by first considering a free-body diagram that includes the sliding support and cuts through both the tie and the arch. Because the compression force in the arch is already known, two unknowns remain: the force transmitted to the support, whose direction must be perpendicular to the plane of sliding, and the internal force in the tie, whose line of action should correspond to the axis of the tie itself.

Were we to start with the free body that includes the fixed support, the problem could not be solved, due to the presence of three unknowns: the internal force in the tie and the two components of the force on the support. This consideration also shows that a system based on two fixed supports must become statically indeterminate. In this case, in fact, we cannot know from which support to begin the analysis.

Design and analysis of arches with ties

Once the stresses in the tie has been found through analysis of the free body that includes the sliding support, the equilibrium of the other free body that includes the fixed support can be determined, so we can immediately find the remaining support force with its two components.

The illustration shows the complete system, with the full construction of the Cremona diagram.

When the line of action of the force on the sliding support is not parallel to that of the resultant of the loads, it becomes possible to directly find the forces on the supports, without having to first find the internal forces in the arch and the tie. We need to isolate a free body that includes the whole structure, and to consider that the three forces that act on the system (the resultant of the loads and the two forces on the supports) are in equilibrium only if their lines of action meet at a single point.

Again in this case, the solution is made possible by the presence of the sliding support, on which the line of action of the force is given (perpendicular to the plane of sliding). If, on the other hand, there were two fixed supports, for which the lines of action of the forces are indefinite, the system could not be resolved in this way because it would be statically indeterminate.

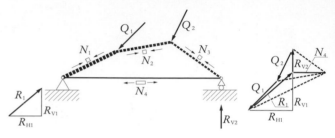

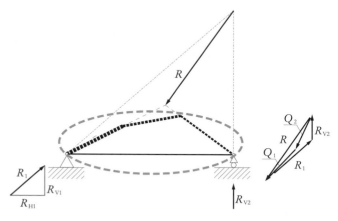

Complete system with the corresponding Cremona diagram

Cable-strut compositions

The new system just described, based on the combination of an arch and a tie, is the result of the need to cancel out or at least to reduce the horizontal thrust component on the supports. We have already met up with the same problem in relation to cables: here too, the supports had to carry both a vertical force and a horizontal one.

The latter could easily be cancelled by attaching the two ends of the cable not directly to the supports, but to an element capable of carrying the horizontal component of the internal force. As shown in the illustration, this element will be subjected to compression: it is known as a strut.

We have already repeatedly emphasized the great similarity between cables and arches; this new structure composed of a cable and a strut is very similar to the arch with tie. In particular, the same considerations apply regarding the support device (one fixed, one sliding, to guarantee a statically determinate state) and the analysis of the internal forces, which can be performed in exactly the same way.

Direct determination of the forces on the supports when the surface of sliding of the mobile support is not perpendicular to the resultant of the loads

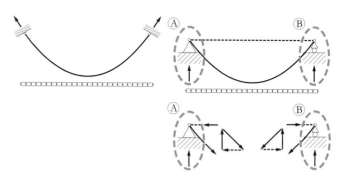

Functioning of a cable with strut compared to a simple cable

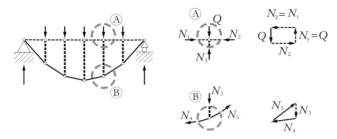

Cable with strut and connection plates

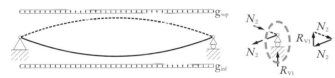

Magazzini Generali in Chiasso, Switzerland, 1924, Eng. R. Maillart

Structure composed of an arch and a cable

The load is first carried by the cable and brought to the supports, where the strut is activated as the contrasting element. Because the loads act only on the upper part of the structure (snow, weight of the roof itself, etc.), secondary elements are necessary, capable of transmitting them directly to the cable, which has the job of carrying them. These elements are under compression and are fully similar to the posts that connect an arch to the deck of a bridge. The cable, which will be loaded with forces concentrated from these posts, will take on a polygonal form.

In the example, we can see the connection posts. In this case the strut adapts to the form of the roof, so it undergoes a deviation halfway along the span. The corresponding force of deviation is transmitted to the cable that also undergoes a deviation in the same direction. The central connection post, in this case, is under tension.

This exemple shows that the strut and the cable can have various shapes, as long as the distance between them be affine to the funicular diagram of the loads.

Arch and cable compositions

It is thus possible to combine arches and cables so that part of the load is carried directly by the arch, while the cable resists the remaining load. The two elements can oppose each other on the supports and transmit only the vertical component to them, when the structure is subjected to loads that are also vertical.

We have already encountered a similar situation when we looked at the roof of the paper mill Burgo by P. L. Nervi (see. p. 48) In this case, we had four load-carrying cables, supported by two pylons and anchored to the roof deck that acts as a compressed strut. As the simplified figure shows, in this case the loads are uniformly distributed and mainly act on the lower, horizontal part of the structure. They must thus be transmitted through the hangers to the load-beaving cables, in a manner analogous to that of a suspension bridge (see p. 47-48).

The fact that the cable remains above the strut in the central part as well depends on the length of the cantilever and the fact that the two pylons that support the cables are inclined in this case. We will investigate in more detain the influence of the ratio between the length of the cantilever and that of the central span on the location of the element in tension when we will look at beams with cantilevers (see p. 176).

Simplified scheme of the paper mill Burgo, Italy, by P. L. Nervi

Cable-stayed systems

In the case of the roof by P. L. Nervi, the roof deck serves as the compression strut, but also as a stiffening girder that ensures a sufficient stiffness under variable loads.

As previously seen on page 59, this problem can also be solved by disposing a large number of cables connecting the strut to a support or a mast. As shown in the figure, this type of structure, termed *cable stayed*, can be considered as the superposition of several tie-strut systems. Technically, the inclined ties are termed stay cables.

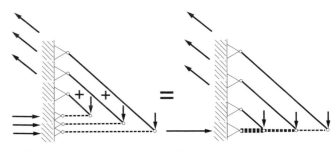
Cable-stayed system as the superposition of several tie-struts

When two cantilevers are disposed symetrically with respect to the support, generally a vertical pylon, the internal forces in the two struts and the horizontal component of the internal force in the stay cables cancel each other out and, under symmetrical vertical loads the only reaction is a vertical force at the bottom of the pylon (see figure). If on the other hand the two cantilevers are not balanced, it is necessary to introduce a supplementary support at one end of the deck to avoid an excessive bending in the pylon.

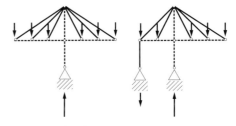
Schemes of cable-stayed systems, with balanced and unbalanced cantilevers

The transborder bridge built by the engineer Arnodin in Marseille is one of the first examples of this structural type. In this case, the stay cables are all connected at the top of the two pylons and constitute thus fans. Since the central span is much larger than the side spans, a truss was added in the middle of the structures, and the whole structure is balanced

Transporter bridge in Marseille, France, 1905, Eng. F. Arnodin

Millau Viaduct, France, 2002-2004, arch. Norman Foster, Eng. Michel Virlogeux

Alamillo Bridge in Seville, Spain, 1991-1992, Arch. Santiago Calatrava, Eng. Jose Ramon Atienza et Carlos Alonso Cobos

by two vertical ties connected to the ends of the stiffening beam and anchored in the ground.

Cable-stayed systems are increasingly used for bridges with an intermediate span (100-800 m) in which the decks work as the compressed strut and the stiffening girder. In the case of the Millau viaduct shown in the picture, the stiffening girder is continuous over several spans and perfectly balanced under permanent loads. Under the action of traffic loads, that can vary from one span to the next, the system is stabilized by the pylons that were designed with a ideal shape for that purpose. It must be observed that the anchorages of the stay cables are not at the extremity of the pylons but are spread at a more or less constant distance over the height of the pylons. This choice results from constructive considerations (dimensions of the anchorages) as well as from their efficiency (length of the stay cables).

It is also possible to dispose the anchorages at regular intervals so that all stay cables are parallel (see example of the Alamillo bridge). This system, called a *harp*, has some advantages and disadvantages over the solution with all stay cables anchored at the top (*fan* configuration). In the first case, the total length of the stay cables is reduced, but the internal forces in the stay cables and the deck is larger, as the figures show.

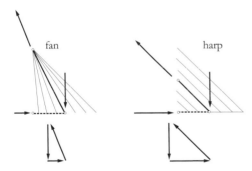

Fan and harp configurations for stay cables, comparison of the resultant of internal forces

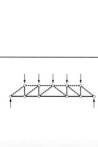

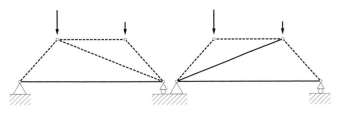

Supplementary bar that opposes the lowering (left) or raising (right) of the arch

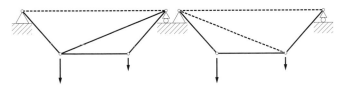

Supplementary bar that opposes the lowering (left) or raising (right) of the cable

Line of action of the internal force for a bar under tension and a bar in compression when the presence of hinges at the ends is guaranteed and the loads act only on the nodes

We have already seen in the previous chapter that the problem of deformability under variable laods and of stability of the element in compression can be solved by adding elements connecting the member in compression to the member in tension. One solution we might want to reconsider and explore further is that of the addition of stabilizing bars (see p. 56 for cables, p. 74 for arches). The illustration shows two cases with a supplementary bar as the stabilizing element of an arch. If the new bar opposes the lowering of the arch, it will be under compression. As we have already seen (p. 74) the opposite side tends to rise; so if we put the new bar in this zone, it will be under tension.

In a structure composed of a cable and a strut the situation is reversed: the supplementary bar that opposes the lowering of the cable will be under tension, while it will be compressed if positioned in such a way as to prevent raising.

Trusses

The structures examined here have the symmetrical geometry of arches and cables, while the load and the corresponding funicular polygon are not symmetrical. In other words, we still have arch-cables, because they are subjected to both tension and compression, but they are no longer funicular. To generalize, we can say that structures of this type do not necessarily have to follow the form of the funicular polygon, so the freedom with which we can design them increases significantly. They are capable, in fact, of resisting any configuration of loads as long.

Analysis of trusses

To analyze these structures, let us take the hypothesis that the nodes function like hinges: the bars are capable of rotating around these points, so the line of action of the internal forces must pass through them. If the bars are straight between one node and the next, and the loads act only on the nodes, the line of action of the internal force must necessarily pass from the two nodes to the extremities and will therefore coincide with the axis of the bar.

Actually, structures of this type are not always constructed with true hinges at the nodes; the bars are often rigidly connected to each other and the lods do not always act on the nodes. The hypothesis we have formulated, then, is not satisfied, but the resulting difference in the internal forces is almost always negligible. In other words, the deviation of the line of action of the internal force from the axis of the bar is minimal.

At this point analysis of the internal forces can begin by considering the equilibrium of a free body that includes one node and cuts through all the bars that connect to it. We have already seen that the problem can be solved only if we do not have more than two unknowns. So we have to begin with a node that connects no more than two bars and on which known external forces act. If the node corresponds to a support, we have to first determine the forces transmitted by the supports themselves, with the method illustrated below.

In the case shown in the illustration, we can begin with a free body that includes the node at the upper left and cuts two of the bars that originally composed the arch. It should come as no surprise that, on analyzing the Cremona diagram and inserting the internal forces thus obtained in the free body, we obtain compressive forces.

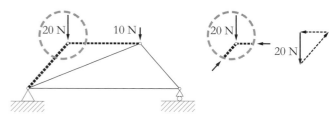

First step of the analysis: free body including only two unknowns

We can now analyze a free body that includes the node in the upper right. Of the three intersecting bars, one now has a known internal force, so the problem can be solved. Completing the Cremona diagram and intbarucing the forces into the free body, we can see that the diagonal bar we have intbaruced is under tension. This confirms our intuitive surmise: the bar that opposes raising must be under tension.

The next free body has to involve a support. The one on the left has three unknowns: it includes a new bar (the tie) and the force transmitted by the fixed support with its two possible components. The free body including the sliding support, on the other hand, has only two unknowns: the internal force in the tie and the force transmitted by the support, which must necessarily be perpendicular to the plane of sliding. So we are forced to analyze the latter free body first.

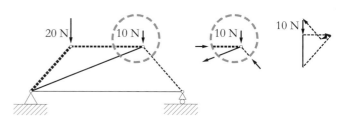

Second step of the analysis: free body including three bars, one of which has already been considered

With the determination of the internal force in the tie, we now know the forces in all the bars. The functioning of the structure is now clear in the diagram in which we have represented, with different lines, the tension and the compression, and chosen the thickness proportional to the intensity of the internal force. At this point the structure can be dimensioned, comparing the internal forces with the specific strength of the material to be used.

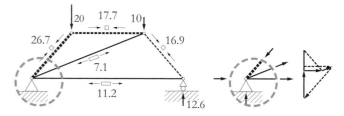

Third step of the analysis: free body with the internal force of a new bar and the force transmitted by the sliding support as the unknowns

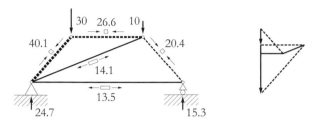

Internal forces with the load to the left increased

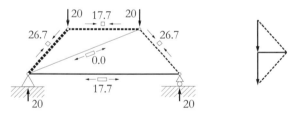

Situation with the funicular polygon of the loads resembling the form of the arch

To complete our analysis of the system, all that remains is to determine the forces transmitted by the fixed support by isolating a free body in this zone. All the internal forces in the bars involved are known, so we only need the force in the support, with its two components.

In this particular case, with all the loads vertical and the sliding support that transmits a force that is also vertical, the horizontal component of the force on the fixed support is necessarily zero. This is also easily seen in the Cremona diagram, where the support force results directly from the closure of the last polygon.

If we modify the acting loads while keeping the same structure, the analysis procedure must be repeated. The illustration shows the final situation of the Cremona diagram and the structure with the complete internal forces for one case of load with the force to the left increased to 30 N. As we can see, all the internal forces have varied with respect to the previous case. Note that we could have obtained the same result by only analyzing the internal forces due to the increase of 10 N of the load to the left, then adding them to the internal forces of the previous case. This principle of superposition is always valid in the case of statically determinate structures with negligible movements and in statically indeterminate structures with linear behavior.

It is interesting to observe how the increase of the load to the left causes a separation of the funicular polygon of loads from the geometry of the arch, so the intensity of the internal forces in the bar we have intbaruced to guarantee stability must necessarily increase.

If we increase the load to the right, instead, to obtain a symmetrical configuration, we will have a funicular polygon that coincides, once again, with the form of the arch. In this case the supplementary bar will simply have no internal force: we have returned to an arch-cable.

**Unstable,
statically
determinate
and statically
indeterminate
systems**

On adding another bar to the structure, we have a situation in which three bars converge in all the nodes. Therefore it is impossible to isolate a free body with just two unknowns: a structure of this type, in fact, is statically indeterminate.

As we have seen, by analyzing a free body that includes a single node on a plane, we can find two unknowns at most: the internal forces in the bars, or reaction forces at the supports. This comes from the fact that all the forces considered already satisfy the condition according to which the lines of action must meet at a single point (the node itself), while analysis of the equililbrium with the Cremona diagram permits us to find only two new unknowns. From this observation, we can deduce a rule that permits us to test whether a system on a plane is unstable, statically determinate or statically indeterminate. Comparing the number of unknowns effectively present with those we can find by using free bodies, we can state that:

- if $n_{reactions} + n_{bars} < 2 \cdot n_{nodes}$ the system is unstable;
- if $n_{reactions} + n_{bars} = 2 \cdot n_{nodes}$ the system is statically determinate;
- if $n_{reactions} + n_{bars} > 2 \cdot n_{nodes}$ the system is statically indeterminate.

Note that by "number of reactions" we mean the number of components of the force transmitted by a support. A sliding support will transmit only a force perpendicular to the plane of sliding: which means that $n_{reactions}=1$. On a fixed support, instead, a force can act with two components, so that $n_{reactions}=2$. If we also block the hinge in a fixed support, so that not only sliding but also rotation is prevented, the unknowns will be the two components of the force and the position of its line of action, which will no longer necessarily pass through the support itself. In this case we have a situation known as clamping, for which $n_{reactions}=3$ (see arches without hinges).

As demonstrated in the examples, this procedure for testing the static determinacy of the system can be used for any type of structure. It is interesting to note that a cable subjected to multiple loads is seen, in this perspective, as a structure with a high degree of instability. In fact, the position of equilibrium is reached only after movements that can be quite large. In the case of the arch with more than three hinges, this situation will clearly lead to instability.

The inequalities described not only enable us to test whether a structure is unstable, isostatic or statically indeterminate, but also represent a valid tool for the design of a stable structure.

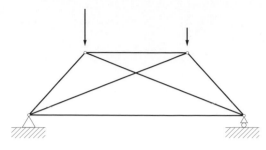

Arch-cable, with two supplementary bars: statically indeterminate system

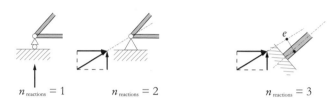

$n_{reactions} = 1$ $n_{reactions} = 2$ $n_{reactions} = 3$

Unknowns found in a sliding support, a fixed support and a clamping

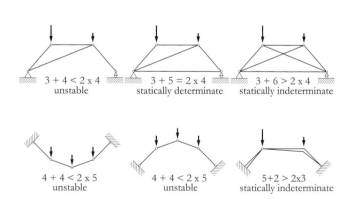

$3 + 4 < 2 \times 4$
unstable

$3 + 5 = 2 \times 4$
statically determinate

$3 + 6 > 2 \times 4$
statically indeterminate

$4 + 4 < 2 \times 5$
unstable

$4 + 4 < 2 \times 5$
unstable

$5 + 2 > 2 \times 3$
statically indeterminate

Static determinacy test for different structural systems

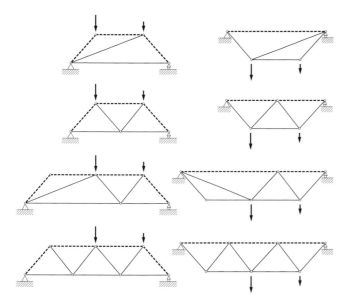

Trusses

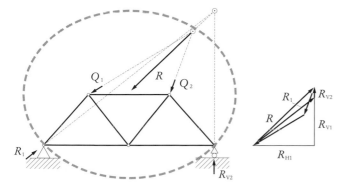

Footbridge on the Michelotti canal in Turin, Italy, 1884, Eng. G. Eiffel

Free body that includes the whole structure

Let us return to the structures composed of an arch, a tie and a single added bar necessary to guarantee stability. We have seen that with three unknowns on the supports, five bars and four nodes, the condition of static determinacy $n_{reactions} + n_{bars} = 2 \cdot n_{nodes}$ is met. If we want to add other bars without threatening the static determinacy, we have to intbaruce a new node for every two supplementary bars. In the first structure shown here we can intbaruce a new node halfway along the tie, attach the stabilizing bar and add a new bar. If we consider the fact that the tie is now composed of two bars, we can see that the principle of static determinacy is met once again.

We can modify the structure again, adding a node and two bars, this time above, in the position of the arch. Clearly this procedure can be repeated infinitely, always obtaining statically determinate structures.

A structure composed of a cable, a strut and a stabilizing bar can also undergo the same metamorphosis and generate a series of systems that are all statically determinate.

Observing the new structures thus generated and comparing them with the originals, we can see a particular feature they have in common: the bars always form triangles placed side by side. These structures are called *reticular systems* or *trusses*.

Intbaruced toward the 16th century as carrying systems for bridges and roofs in wood, trusses were greatly developed in the 19th century, through the use of steel, thanks to their remarkable efficiency.

Generation of trusses

The procedure described above for the analysis of a simple truss is clearly applicable to all structures of this type. As we have seen, the first step is to analyze a free body containing only two unknowns. Often the nodes in which just two bars converge correspond to the supports. In these cases it is first necessary to determine the forces transmitted by the supports themselves. As we have seen in the case of cable-arches, when the resultant of the loads is not parallel to the force transmitted by the sliding support, it is possible to utilize a free body that includes the whole structure, and to determine the support forces, considering the fact that the lines of action of the three forces involved must meet at a single point. Note that in this case, analyzing a single free body, we can directly find not two but three un-knowns. This is made possible by the fact that, in this case, we can also make use of the condition that links the lines of action. This allows us to determine the direction of the force at the fixed

General analysis of trusses

support, reducing the latter to just one unknown (intensity of force with known direction).

Instead, when the resultant of the loads is parallel to the force of the sliding support, the meeting point of the three lines of action is infinity, so the procedure described can no longer be used.

In this case the analytical solution of the problem represents a valid alternative (see appendix 2, p. 233). If we want to continue to use the tools of graphic statics, we can replace the truss with a arch-cable and determine the forces on the supports with the method employed for those structures. This is made possible by the property of statically determinate systems: with equal loads and configuration of the supports, the intensity and the direction of the forces transmitted by the supports is independent of the type of structure. To better understand this property, we can recall how the forces on the supports can be found by isolating a free body that includes the entire structure. The type of structure enclosed in the free body, then, has no influence on the unknowns being sought.

The first step, then, is to construct a funicular arch (or a cable) with the familiar procedure: determination of the line of action of the resultant by means of an auxiliary arch and construction of the funicular arch, considering that the two bars near the supports, if lengthened, must meet on the line of action of the resultant. To permit easy comparison of the functioning of the funicular arch-cable and that of the truss we want to analyze, we select the rise of the funicular arch as identical to that of the truss. For the case we are analyzing, this means that the internal forces in the bars, near the fixed support, are identical in the two structural systems.

We can then determine the forces transmitted by the supports, isolating the free bodies that include them. First we consider the sliding support, in which the unknowns are the internal force in the tie and the force perpendicular to the plane of sliding. Then we complete the procedure by analyzing the fixed support, in which the unknowns are reduced to the two components of the force, because the internal force in the tie is now known.

Once the forces transmitted at the supports have been found (15 N from the sliding support, 25 N from the fixed), we can intbaruce them into the truss we want to analyze and begin our study of the internal forces, observing one of the two free bodies that include a support. Because the forces on the supports are now known, both these free bodies have only two unknowns, so we can begin with either one of them.

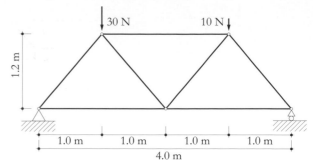

Truss with the resultant of the loads parallel to the force on the sliding support

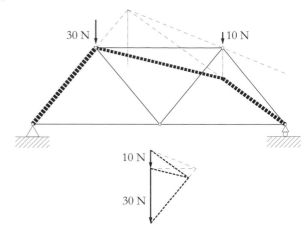

First step: construction of a funicular arch

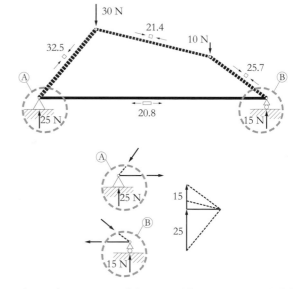

Second step: determination of the internal force in the tie and of the forces transmitted by the supports

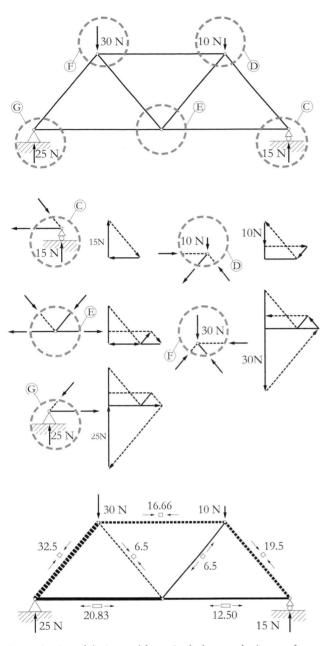

Determination of the internal forces in the bars, analyzing one free body at a time

In the solution shown here, first we consider the sliding support with the free body C that permits us to determine the internal forces in the tie to the right and in the last bar of the arch. Note that the stress in this part of the tie does not correspond to the one previously found for the arch-cable.

We can then move on to free body free body D, which includes the node to the upper right, and so on, isolating one free body at a time to determine all the internal forces in the structure. Finally, we analyze free body G, which includes the fixed support. In this case we have no more new unknowns, so we can use this free body to test our work.

To facilitate comprehension, we have redrawn and completed the adjacent Cremona diagram each time we have passed to a new free body. Clearly, both to save time and to limit the inaccuracies caused by repetition of the drawing, it would be better to construct only one Cremona diagram.

The last figure, with the graphics showing the tensile and compressive internal forces and the width of the lines in proportion to the intensity of the internal force, permits us to visualize the functioning of the structure. For example, we can still recognize that the arch is under compression and the tie under tension. Note, in the latter, that the intensity of the tension varies along the length.

Upper chord, lower chord and diagonals

We have constructed trusses by combining arches, ties, cables, struts and supplementary bars that stabilize the structure and make it statically determinate. This approach is very useful to apply what we have learned about cables and arches, for a better understanding of the functioning of trusses.

But in our discussion of trusses we also need to intbaruce some new terms: the upper bars of the arch are part of the so-called *upper chord*. In like manner, our tie is defined as the *lower chord*, while the bars that connect the two chords are known as *diagonals*.

The distance between the lower and upper chords, corresponding to the rise of the arch, is defined as the depth of the truss. When the chords are parallel, we say we have a truss of constant depth.

Influence of the height and the span on the internal forces in trusses

We have already seen, in the case of cables and arches subject to vertical loads, how the horizontal component of the internal forces is directly proportional to the span and inversely proportional to the rise. In other words, by doubling the span or halving the rise, the internal forces double.

For trusses, the same relationships apply between internal forces in the chords, depth and span. The illustrations show the analysis of two trusses, similar to the one just described, in which the depth is doubled or halved. From the Cremona diagram based on the truss with double depth, it is clear that all the horizontal components of the internal forces have been halved. Similarly, by halving the depth, we obtain double internal forces in the chords.

We should recall that, for equal loads, the Cremona diagram depends only on the slope of the various bars, while it is independent of the absolute dimensions. For this reason, the halving of the depth, in our case, has the same effect as the doubling of the span on the internal forces. The internal forces in a truss with a span $\ell=4.00$ and height $h=0.60$ are therefore identical to those of a truss subjected to the same loads, with measurements of $\ell=8.00$ and $h=1.20$ m.

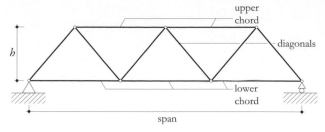

Truss with upper chord, lower chord and diagonals

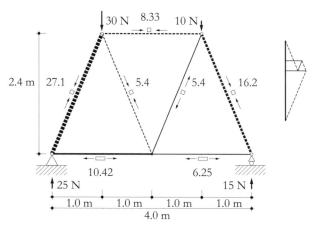

Truss with doubled height

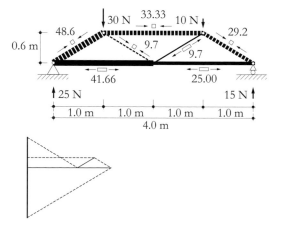

Truss with halved depth (or doubled span)

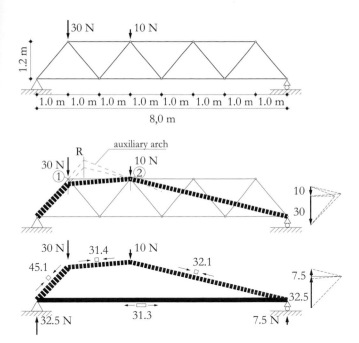

Truss with parallel chords and funicular arch with tie

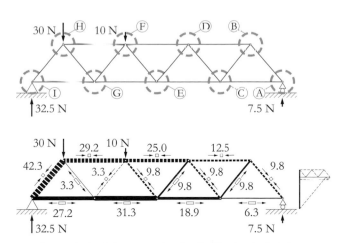

Complete analysis of the truss using free bodies that include just one node

Complete analysis of a truss

We can proceed with the analysis of a truss with any number of bars using the same procedure. The sole condition is the static determinacy of the structure. As we have seen, we have to begin our analysis with a node that has only two bars, in order to have just two unknown quantities. When this node corresponds to a support and the resultant of the loads is parallel to the forces transmitted by the supports, it is first necessary to determine these forces by means of an auxiliary arch-cable.

Later we will see how this easily constructed static system can become very useful for understanding of the functioning of the truss. Therefore its construction is recommended, even when it is not indispensable for the analysis of the truss (nodes with two bars that do not correspond to the supports, or the resultant of the loads not parallel to the forces at the supports).

In the example shown here, we have chosen a funicular arch with a rise identical to the depth of the truss. This permits direct comparison of the internal forces in the two systems. Note that the maximum rise of the arch is reached on the line of action of the load of 10 N and not, as in the previous example, in correspondence to the other load of 30 N. If we want to construct an arch with the rise equal to the depth of the truss, then we have to draw a segment of the arch passing through the support to the right and node 2. Extending this arch as far as the line of action of the resultant, we obtain a point, which when it is connected to the left support permits us to construct the last segment of the arch.

Once the forces that sustain the structure through the supports have been determined, we can begin to determine the internal forces in the bars, analyzing one free body after another. For our truss with nine nodes, then, we will have to analyze nine free bodies. In fact, we could analyze just eight, because the last one contains no new unknowns, so its analysis simply represents a way of checking.

The result of the analysis is summed up in the diagram in which the bars are represented with a width proportional to their internal forces, as if the bars had been effectively dimensioned ($A = N_d/f_d$).

It is interesting to observe that the internal forces in the four bars that form the lower chord undergo sizeable variations, from a minimum of 6.3 N near the right support, to a maximum of 31.3 N in the central zone (see the diagram of internal forces repbaruced here).

If we compare these internal forces with that of the tie in the arch-cable, we see that the maximum intensity coincides but

TRUSSES

133

that the shape is quite different (constant across the entire length of the tie). Instead, there is great similarity between the drawing of the internal forces in the chords and the form of the funicular arch.

To understand this situation, which is certainly not random, we have to go back to the relationship that exists between the depth h of the truss and the internal force N in the chords: the two quantities are inversely proportional (by doubling h we obtain a halving of N, and vice versa).

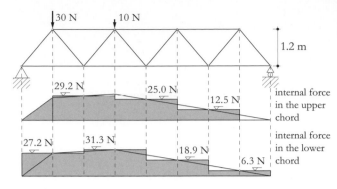

Distribution of internal forces in the upper and lower chords

Bending moments

This is the same as saying that the pbaruct of h and N is independent of the chosen truss. In fact, this quantity, measured in Nm and defined as the bending moment in statics, depends only on the loads and the span, while it remains the same if the structure is varied. For the truss with parallel chords, the arch-cable (the funicular arch with tie, for example), and even for any other type of structure with the same loads and the same supports, we will therefore have an equal bending moment (intensity and shape are identical along the whole length of the structure). The two structures analyzed are, in effect, two particular cases: in the first, the depth h is constant, so the internal force N follows the shape of the bending moment, while in the second case, as the internal force N of the tie is constant, we have a shape of the depth h similar to the pattern of the bending moment .

In the statics, procedures have been developed that permit calculation of the bending moment at any point in the structure, without requiring the construction of the funicular arch and determination of the internal forces in the chords (see appendix 3, p. 235). This approach, more closely linked to the needs of the engineer, is different from the one we will examine here. Later, we will utilize the properties of the funicular arch to ease understanding of the structures we discuss, however we will overlook the use of the bending moment.

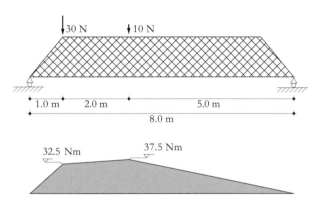

Shape of the pbaruct $N \cdot h$ (bending moment) for the truss, the arch with tie, and any structure with the same loads and the same supports

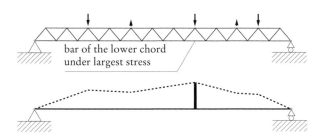

bar of the lower chord
under largest stress

Identification of the bars under largest stress in the chords by means of
the analogy with the funicular arch

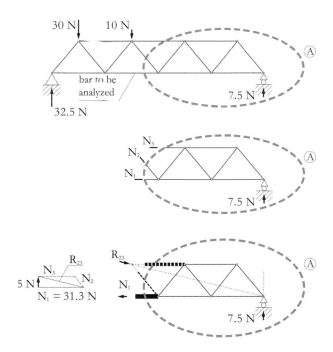

30 N   10 N

bar to be
analyzed

32.5 N

7.5 N

N_3

N_2

N_1

7.5 N

R_{23} R_{23}

N_3

5 N N_2 N_1

$N_1 = 31.3$ N

7.5 N

Direct determination of the stress in any bar

The resemblance between the form of the funicular arch and the shape of the internal forces in the parallel chords of a truss is useful to find which bars are most stressed in the chords. Once we have established where the funicular arch has its largest height, we can immediately identify these bars. This possibility becomes very useful when the truss is composed of many bars and a complete analysis would require laborious examination of very many free bodies (see the example).

When the chords are not parallel (variable truss depth) the identification of the bars with the largest internal force becomes more complex. We have to find the bars for which the ratio between the height of the funicular arch and the actual depth of the truss is largest.

Once the bars with the largest internal force have been identified, and after finding the forces at the supports, it is possible to directly determine the intensity of its internal force by means of a properly selected free body. The free body must necessarily intersect the bar we want to analyze.

The free body A shown here satisfies this requirement: the internal force in the most stressed bar acts on it, as well as that of two other bars and the force transmitted by the sliding support. If the latter quantity has already been found, only three unknowns remain. In this case the solution is possible if we take the two conditions of equilibrium into account: the forces must vectorially cancel each other out, and the lines of action of the three forces in equilibrium must meet at one point. Given the fact that four forces are involved, of which one is already known, we have to first sum up two of them in a partial resultant in order to have, in an initial phase, just three forces with three lines of action. We can, for example, consider the partial resultant R_{23} of the two internal forces N_2 and N_3, which at the moment do not interest us. Its line of action must pass through the node where the two bars meet and, at the same time, intersect with the internal force N_1 and of the force on the support. Once this line of action has been defined, the problem is reduced to finding two unknowns (N_1 and R_{23}) with known lines of action.

The force polygon that permits us to immediately find the two unknowns, including the internal force in the bar under the largest stress in the lower chord, can eventually be completed, breaking down the partial resultant R_{23} in such a way as to also determine the internal forces in the diagonal and the upper chord (N_2 and N_3).

Identification of the bars under largest stress in the chords

Targeted analysis of bars in the chords of trusses

Note that the resultant R_{23} corresponds to the internal force we would have found had we chosen free body B. Like free body A, it intersects the bar we want to analyze but, instead of intersecting the diagonal and the upper chord, it cuts the node. This means that the partial resultant R_{23} is simply the internal force present in the node. Analysis of free body B, then, is the most direct way to determine the force the bar that interests us.

When multiple external forces act on a free body, as in the case of free body C shown here, they must first be added together in order to have only three forces to analyze: the resultant R_Q of the external forces, the internal force N_1 of the bar we want to determine, and the internal force R_n in the node (or the resultant of the internal forces in the other two bars if the free body intersects them instead of the node).

The resultant R_Q of the external forces that act on the free body, in practice the loads and the forces at the supports, can be determined with one of the methods we already know: vectorial sum of the forces and construction of the line of action passing through the point of intersection of the lines of action of the loads when the forces are not parallel, or determination of the line of action of the resultant by means of an auxiliary arch or cable when the forces all act in the same direction.

Note that the funicular arch with tie we used to identify the bars with the largest internal force can also be seen as an auxiliary structure that permits us to find the line of action of the resultant. The line can be determined by analyzing the same free body that includes, this time, a part of the funicular arch and the tie. We only have to lengthen the intersected bars until we find the point of intersection through which the line of action of the resultant must pass.

The procedure shown in the illustration makes it clear that, in this case, the internal force R_n in the node corresponds precisely to the stress in the funicular arch when the rise at the node is identical to the depth of the truss. This applies, in fact, only when the node is on the upper chord.

We should also notice that the internal force R_n in the node we have just determined refers to the same node examined previously (see p. 135). Yet the internal forces do not correspond because the two free bodies cut the node in different ways: free body B to the right of the external force, free body C to the left. This difference is evident when we observe the funicular arch as well: the internal force and, above all, its slope vary, shifting from one part to the other of the point of application of the force.

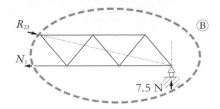

Free body that intersects a node and a bar of a chord

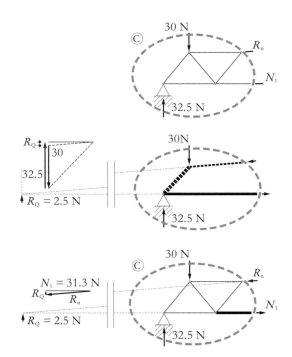

Free body with multiple external forces, determination of their resultant by means of the arch with tie

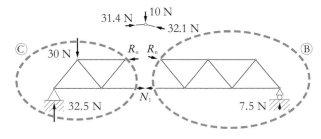

Internal forces in the nodes and stress in the funicular arch

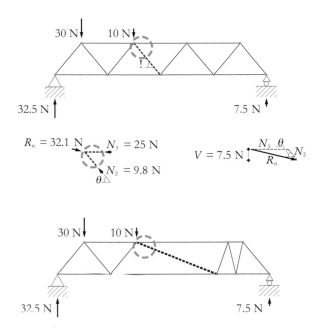

Internal force in the diagonals and the influence of their slope

As we have already seen (p. 135), the procedure that permitted us to find the internal force in the chord can be completed by determining the internal force in the diagonal and that in the other chord included in the free body. This happens by breaking down the internal force R_n in the node into two directions: that of the chord and that of the diagonal.

In a truss with horizontal chords, the diagonal has the function of carrying all of the vertical component of the internal force in the node. This fact makes it clear that the internal force in the diagonal also depends on its slope. The illustration shows how a diagonal with less slope has a larger internal force.

When the loads are vertical, the vertical component of the internal force in the node R_n is identical to the resultant of the loads on the free body R_Q (see previous section). As in the case of the bending moment, it depends, then, only on the loads and the position of the supports. In the statics this force is defined as shear force (V). If we find this value, we can immediately find the internal force in the diagonal based on its slope θ:

$$N_2 = V / \sin(\theta)$$

When the slope θ is very small the internal forces will be very large. For this reason, the diagonals are usually arranged in sufficient numbers to guarantee a relatively large slope and therefore to limit their internal forces.

Analysis of the diagonals and their functioning

Shear force

Identification of diagonals with the largest internal force

As we have just seen, the vertical component of the internal force R_n in the node, the vertical component of the stress in the funicular arch and the shear internal force V are actually one and the same and differ only in the approach used. From this consideration, we can deduce that the internal shear force V is proportional to the slope of the funicular arch. When the diagonals have equal slope, their internal force will therefore be greater where the funicular arch is most inclined.

In the trusses we have analyzed thus far, with two supports at their ends, subjected only to downward loads, the diagonals with the largest internal forces are always located at the supports. In fact, in these zones the funicular arch has a steeper slope.

Study of the slope of the funicular arch is very useful in the case of complex systems or loads directed both downward and upward.

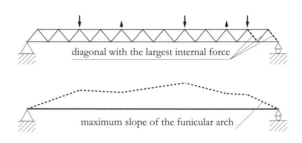

Identification of the diagonals with the largest internal force by using the funicular arch

Identification of diagonals under tension and those under compression

The illustration shows the analyses of two diagonals: the first is under compression, while the second is under tension. In the first case the stress in node R_n on the upper chord corresponds to the internal force in the funicular arch and pushes the free body downward. The diagonal opposes this, and therefore will be under compression.

The next diagonal can be analyzed by considering a node located on the lower chord. In this case the internal force R_n no longer corresponds to the stress in the funicular arch (we would have to intbaruce a strut and a cable, then the internal force of the latter would correspond, once again, to R_n). In any case, its vertical component is in equililbrium with the resultant R_Q of the loads and still exerts a downward pressure on the free body. Breaking down this resultant into the direction of the lower chord and that of the diagonal, we can find the tensile force in the diagonal.

To generalize and simplify these results, we can say that the diagonals sloping in the same direction as the funicular arch will be under compression, while those sloping in the opposite direction will be under tension. In fact, this is a simplification that is only valid when the chords are parallel, but it is very useful to distinguish between the diagonals under tension and those under compression, simply with the help of the funicular arch.

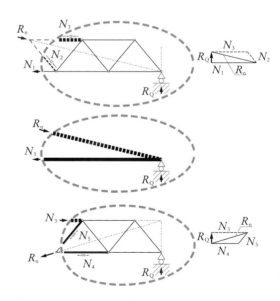

Diagonals under compression and tension

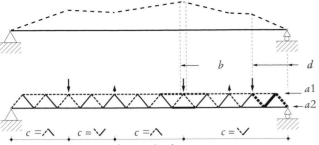

a1: upper arch compressed upper chord
a2: lower tie lower chord under tension
b: maximum distance between arch and tie maximum stress in the chords
c: diagonals under tension or compression
d: maximum slope of funicular arch: most stressed diagonals

Qualitative analysis of the truss by analogy with the arch-cable

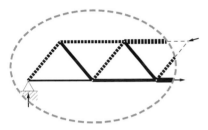

V diagonals (Warren system)

Centre Georges Pompidou in Paris, 1977, Arch. R. Piano and
R. Rogers, Eng. P. Rice (Ove Arup & Partners), reticular beam with
V diagonals during assembly (ℓ = 44.8 m, h = 2.85 m, ℓ/h = 15.7) and
view of the trusses on the facade (the bars that connect
the reticular beams are there to stabilize the entire construction)

Thus we may consider the arch-cable as a structure that is easy to analyze, also in quantitative terms, and it allows us to study the quality of a truss with parallel chords. The comparison of these two structures can help us to formulate four very useful rules:

    a. when the arch is above the tie in the arch-cable, the upper chord of the truss is under compression, while the lower chord is under tension;

    b. where the distance between the arch and the tie is largest, we find the bars of the chords of the truss with the largest internal force;

    c. the diagonals of the truss that slope in the same direction as the funicular arch are compressed, while the others are under tension;

    d. for equal slope, the diagonals of the truss with the largest internal force will be located where the funicular arch is most inclined.

Qualitative analysis of a truss

The trusses described thus far are characterized by diagonals arranged in such a way as to form a series of V shapes. This is why they are called "trusses with V diagonals" or "Warren trusses", from the name of the engineer who patented the system in 1848.

Possible configurations of the diagonals

The illustration shows an example of a truss with V diagonals in which the bars are constructed in response to the type of internal force: all the bars in compression have a tubular section, while all the bars under tension are solid and thus have a smaller diameter. Later, discussing problems of stability, we will see that steel sections under compression, in fact, must have a very definite form, while bars under tension can have any section. In general, we can say that for bars under compression the most efficient sections are tubular or double-T, while for bars under tension, flat or round sections can also be used.

V diagonals

For this reason, the qualitative analysis of trusses we have described is very useful in the design of truss structures.

N diagonals

One variation of a truss with V diagonals can be pbaruced by diminishing the slope of the compressed diagonals and increasing that of the diagonals under tension, to the point of making them vertical (in this case they are called *posts*). The result is the so-called "truss with N diagonals under compression".

These trusses are often known as the "Howe system" (W. Howe patented it in 1840), although a structure of this type had already been proposed by Andrea Palladio almost three centuries earlier.

With vertical posts, the rule that links the type of stress to the slope of the bar, in comparison to the slope of the funicular arch, is no longer valid. By isolating a node that connects the post to the diagonal and the chord, it is easy to demonstrate that in this type of truss the vertical bars are generally under tension.

If instead we increase the slope of the diagonals under compression, under until they are transformed into posts, we obtain a "truss with N diagonals under tension", also known as the "Pratt system" (C. Pratt, patent 1844).

X diagonals

By overlaying a truss with N diagonals under tension and another with N diagonals under compression, we obtain another configuration of the diagonals. This is a truss with posts and diagonals in the form of an X, also known as the "Long system" (H. Long, patents in 1830 and 1839). The new structure is clearly statically indeterminate: for every field there is a surplus diagonal.

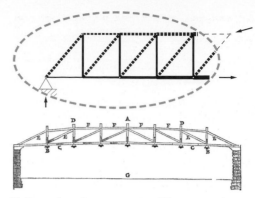

N diagonals under compression (Howe system), structure proposed by A. Palladio in 1570 (*The Four Books of Architecture*, book III, chapter 8)

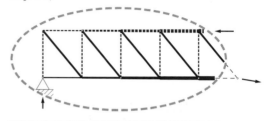

N diagonals under tension (Pratt system), design by Mies Van der Rohe for the National Theater of Mannheim, Germany, 1953 ($\ell = 80$ m, $h = 8$ m, $\ell/h = 10$)

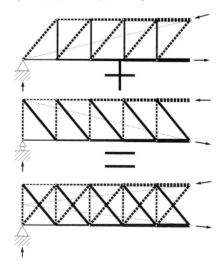

X diagonals with posts, obtained by overlaying N diagonals under tension and N diagonals under compression

Reticular beams of the Crystal Palace in London, UK, 1851, Arch. J. Paxton, Eng. Fox Henderson (ℓ = 7.32, 14.64 and 21.96 m, h = 0.915 m, ℓ/h = 8, 16 and 24)

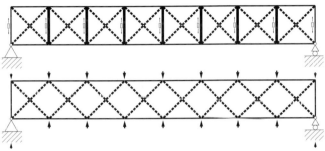

Internal forces in a statically indeterminate truss with X diagonals, due to the effect of pre-tensioning of the posts; statically determinate truss without posts with external forces on the nodes, to simulate the effect of pretensioning of the posts

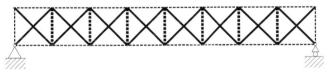

Truss with X diagonals and posts, in which the diagonals are pretensioned

Wright brothers' biplane, first flight on 17.12.1903

The effect of an external action is to activate the two original systems: one part of the load is carried by the truss with N diagonals under tension, while the remaining part is carried by the other truss with N diagonals under compression. The posts, which perform their function for both systems, are subjected to compression in the first case, and to tension in the second. These internal forces compensate for each other, so the internal force in the post is usually very small with respect to that in the diagonals.

The strong static indeterminacy of the system permits the intbaruction of pretensioning internal forces in the bars. These are internal forces that result from lengthening imposed on certain bars before their installation, and they are completely independent of the external loads.

We can imagine a truss with X diagonals without any external load, in which the posts are too short and are lengthened by means of tension before being positioned at their corresponding nodes. This internal force is opposed by the diagonals, which in this case are subjected to compression, and by the lower and upper chords that are subjected to tension. These internal forces are identical to those that would exist in a statically determinate truss without posts, loaded with downward vertical forces on the nodes of the upper chord, and upward forces on those of the lower chord.

If the pretensioning internal force in the posts is always greater than that of the compression resulting from external loads (both permanent and variable), the posts will always be under tension and can be composed of metal cables.

A truss in which the posts are prestressed (shortened by means of compression before attachment to the nodes, being initially too long) will have the reverse situation, with the diagonals under tension and the chords under compression. Note that the same result can be achieved through pretensioning of the ties.

This principle was often used in the first biplane aircraft, in which the two wings formed the chords and the pretensioned diagonals were composed of very slender steel cables. The posts, always under compression, were instead formed by wooden bars.

Because in a truss with bars resistant to both tension and compression the posts can be considered to be surplus, we can imagine a truss with X diagonals and no vertical bars. A structure of this type can also be obtained by stacking two trusses with V diagonals in a staggered way.

If posts are to be included at the two supports, the system will be statically indeterminate to just the first degree.

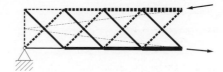

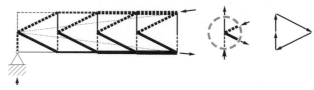

Truss with X diagonals, without posts

K diagonals

Another configuration of diagonals and posts is shown here: that of "trusses with K diagonals".

The diagonals are connected to nodes at the center of the posts, which thus divided are under compression in the lower part and under tension in the upper part. The upper diagonal, sloping in the same direction as the funicular arch, is under compression, while the lower diagonal, with its inverse slope is under tension. This can easily be demonstrated by isolating a free body that intersects the two diagonals and the two posts.

This system is statically determinate. In fact, it is possible to begin its analysis with the node corresponding to the support and to treat all the other nodes, one at a time, using free bodies, without ever having more than two unknowns at a time.

It is possible to invert the two diagonals of the previous system to obtain tension in the upper diagonal and compression in the lower. This configuration is very similar to the previous one, yet it is less efficient, because it creates larger internal forces in the chords.

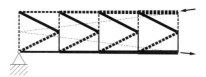

Truss with K diagonals

Truss with K diagonals, in the opposite direction with respect to the funicular structure

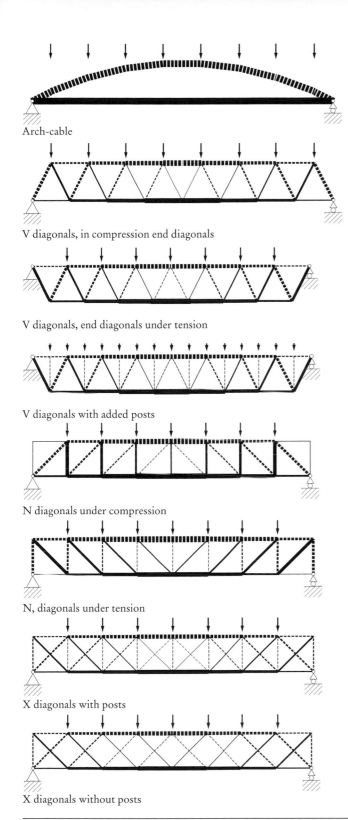

Arch-cable

V diagonals, in compression end diagonals

V diagonals, end diagonals under tension

V diagonals with added posts

N diagonals under compression

N, diagonals under tension

X diagonals with posts

X diagonals without posts

The illustrations show the most common configurations, comparing them with the arch-cable. The loads always have the same intensity and the internal forces are always represented in the same way (thickness proportional to internal force).

As indicated in these figures, in trusses with V diagonals the position of the supports (on the upper or lower chord) influences the stress in the diagonals at the ends.

The insertion of added posts in a truss with V diagonals is usually required when the load acts directly on the upper chord or when the upper chord needs to be stabilized. If strong loads act on the lower chord it is necessary to insert posts that suspend it from the nodes of the upper chord.

As we have seen, trusses with X diagonals and posts have marked static indeterminacy (the example shown has a degree of static indeterminacy equal to eight). We should notice that here the diagonals are subjecter to lower internal forces than in the previous cases.

Trusses with X diagonals but without posts are also statically indeterminate, but they have just one surplus bar.

TRUSSES

Trusses with K diagonals are more efficient when the two diagonals (upper and lower) converge on the same side as the funicular arch-cable composed of arch and tie or cable and strut.

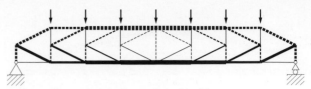

K diagonals, opening similar to the arch-cable

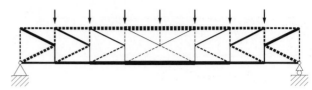

K diagonals, opening inverted with respect to the arch-cable

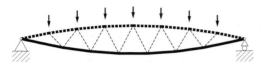

Lenticular truss corresponding to the funicular structure in the case of distributed loads

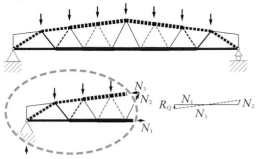

Bridge on the Isar at Grosshesslohe, Germany, 1857,
Eng. F.A. von Pauli, H. Gerber and L. Werder

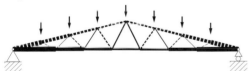

Trapezoidal truss, internal forces with distributed loads

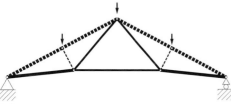

Triangular truss subjected to uniform loading

Polonceau system truss

Roof of the Gare du Nord in Paris, France, 1865, Arch. J.-I. Hittorff,
system invented by C. Polonceau in approx. 1840

Forms of trusses

Until now we have examined two particular cases: trusses of constant depth with parallel chords, and funicular structures composed of arches, cables, ties and struts in which the shape of the depth corresponds to the shape of the funicular polygon of the loads. As we have seen, in funicular structures the diagonals, if present, are not subjected to internal forces or transmit only the force of deviation of the internal force in the chords at the point of application of the load, as seen in the example shown here. On applying a variable load, there will obviously be a modification of the funicular polygon, so the diagonals will also be activated.

Often the form of trusses responds to other requirements. The trapezoidal truss illustrated here, for example, follows the shape of the roof. In these trusses, if the loads are evenly distributed, the central part is usually marked by internal forces in the diagonals that are opposite to those of trusses of constant depth. This is due to the fact that in this zone the internal shear force caused by the loads is smaller than the vertical component of the internal force in the upper chord. The maximum internal forces in the chords, furthermore, are reached not at mid-span (the point in which the truss reaches its maximum height) but in the intermediate zones where the ratio between the depth of the funicular polygon and that of the truss is largest.

One extreme case is that of triangular trusses, in which all the internal forces in the diagonals are inverted with respect to those of trusses of constant depth. Moreover, the maximum internal force in the chords is reached near the supports.

The structure shown here, known as the Polonceau system, is a particular case of a triangular truss. It can be seen as the composition of two inclined trusses arranged in an arch and connected with a tie, to compensate the thrust. Again in this case, the bars with the largest internal force are those near the supports. This solution was widely used in the second half of the 19th century for the construction of railroad stations. Thanks to the fact that the lower visible elements are above all under tension, and can thus be formed by slender cables, these structures are usually largely transparent.

Form and structural efficiency

The form of the truss, seen as the form of the chords and the configuration of the diagonals, and above all the slenderness of the structure (ratio between span s and depth h), have a decisive influence on the internal forces in the bars and, therefore, also on the quantity of material required to resist these internal forces.

The diagram shown here, very similar to the one seen earlier for cables (p. 50), shows the quantity of material required in keeping with the slenderness ratio ℓ/h. It includes all the structures studied thus far: arches and cables, arch-cables, and trusses of constant height with N, V, X and K diagonals. Only structures loaded with a force concentrated at mid-span and supported on two supports at the end are considered here. To facilitate comparison of these different structures, the load is always applied at the level of the supports (the arches and cables are therefore completed with supplementary vertical bars). Moreover stability is not considered, so the necessary section comes from the equation $A_{nec} = N_d/f_d$. As we will soon see, the bars in compression that exceed a critical slenderness require, to avoid instability, a larger area than the one indicated by the design equation. Furthermore, the material considered has the same resistance, both to tension and to compression.

In trusses of constant depth there is a sizeable variation in the internal forces in the chords, so the optimal structure has variable sections. Often, for practical and aesthetic reasons, this variation is avoided and the chord, designed for the maximum internal force, is therefore oversized through the better part of its length. In the diagram we see the curves that refer both to oversized trusses with a constant section of the chords, and to those with a variable section of the chords to reduce the quantity of material required to a minimum.

We can make the following considerations in light of the diagram:

- For all the structures, beyond a certain limit (ℓ/h of about 3), the increase of the slenderness leads to an important increase in the quantity of material necessary (self-weight of the structure).
- Cables and arches require the same quantity of material. This is evident, from the similarities between these two structural forms; actually, if we consider the problem of the stability of compressed bars, the arch would require a much greater volume of material.

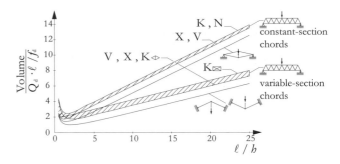

Quantity of material in relation to the slenderness ratio ℓ/h (with equal span ℓ, load Q_d and strength of material f_d)

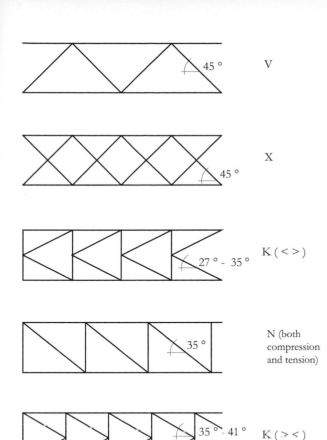

45°   V

45°   X

27° - 35°   K (< >)

35°   N (both compression and tension)

35° - 41°   K (> <)

Slopes of diagonals for which the volume of required material is minimum (trusses of constant depth)

– The arch-cable, which can be seen as a truss whose form corresponds to that of the funicular polygon, requires about twice as much material as an arch or a cable. This comes from the fact that in the arch, or in the cable, the horizontal thrust is carried by the supports, while in the arch-cable the thrust is carried by an element of the structure that requires its own quantity of material.
– For slenderness ratios greater than 5, trusses of constant depth, in which the area of the chords follows the shape of the internal forces, are much more efficient than cable-arches: for chords, a quantity of material comparable to that of the arch and the cable is required (comparable internal force and length), although we must also add material for the diagonals.
– The configuration of the diagonals also has an influence on the volume of material, but it is less important than the other factors. For trusses with chords of variable section the most efficient diagonals are X, V and K (<>), while the N and K (><) diagonals require more material. For trusses with oversized chords of constant section, the situation is similar: X and V are the most efficient, N and K (both <> and ><) are the least efficient.

The slope of the diagonals influences the volume of material too. The diagram covers only the slope that reduces the necessary material to a minimum (see the optimal slopes illustrated here).

Influence of form on structural stiffness

We have often remarked on the fact that a structure should not only be resistant to loads without breaking (the ultimate limit state criterion for design), but should also be capable of deforming without exceeding certain limits (the service limit state criterion). So the efficiency of a structure also depends on its stiffness.

The diagram shows the maximum deflection of the structures previously analyzed, expressed once again in terms of the slenderness ratio. Here we have considered a load that causes a unitary deformation equal to 0.001 in the bars with the largest internal force (for example, stresses $\sigma = 205$ N/mm² and modulus of elasticity of steel $E = 205\,000$ N/mm²). Thanks to the principle of proportionality, valid for structures with linear behavior, the same diagram can be used for other internal forces and other materials with different moduli of elasticity.

Again in this case, an increase in slenderness generally causes a major increase of deflections.

The stiffest structures are trusses of constant depth with chords that have a constant section. This is due to the fact that these structures are oversized, so many bars are not completely exploited and their unitary deformation reaches the value hypothesized only in the zones of largest internal force.

The structures without surplus material and with the smallest deflection are arches and cables (note that the values indicated in the diagram on p. 51 are only larger because a larger internal force and a smaller modulus of elasticity were hypothesized).

Next come trusses of constant height with variable-section chords, with slightly larger deflections. What we have observed regarding the quantity of material also applies here for the influence of the configuration of the diagonals and their slope.

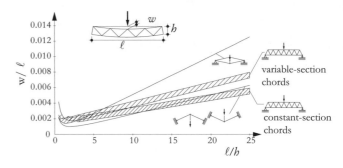

variable-section chords

constant-section chords

Mid-span shift caused by a concentrated load in keeping with the slenderness ratio ℓ/h (unitary deformation $\varepsilon = 0.001$)

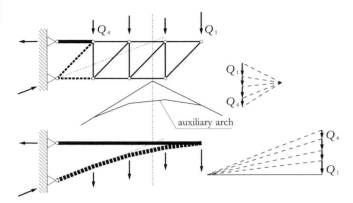

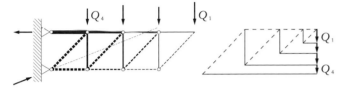

Analysis of a cantilever truss (or tower) using the arch-cable

Complete analysis of the truss with a free body for each node

We have already discussed cantilevers with a single concentrated load at their extremity in the chapter about arch-cables. Indeed, in this case, a tie and a strut are sufficient to carry the load to the supports. When several loads are present, if the selected structural system is a truss, the structures will be as indicated in the figure. This system can be analyzed by repeatedly applying the same procedure:

– determination of the resultant of the loads by means of an auxiliary structure (arch or cable);
– study of a arch-cable (composed of an arch and a tie, a cable and a strut, or an arch and a cable);
– identification of the type of stress (tension or compression) and identification of the bars with the largest internal force by means of the affinity between the truss and the arch-cable;
– determination of the intensity of the internal force in the bars involved, using the appropriate free bodies.

In cantilevers subjected to the force of gravity, the tie is above the arch, so the upper chord will be under tension while the lower will be compressed. The arch reaches its largest depth (distance from the tie) at the connection to the wall (or to the ground, for the tower), so the corresponding bars of trusses with parallel chords will be the ones with the largest internal force.

From the slope of the funicular arch we can deduce that the diagonals will all be under compression, and that those near the wall will be the ones with the largest internal force.

Clearly the possibility remains of examining one node at a time to determine the internal forces in all the bars (see the construction shown in the illustration).

Towers

Towers are subjected to vertical gravity loads and by horizontal loads induced by wind or seismic accelerations. If vertical loads can be carried to the ground by columns and cores only subjected to compression, under horizontal loads, the structure will behave exactly like a cantilever subjected to vertical loads. Indeed, we can imagine the same structures turned 90 degrees and subjected to vertical gravitation forces. The chords of the structure are formed by the columns of the tower, and diagonals need to be inserted to connect them. Since in this case the floors form the posts of the truss, a system with K or X diagonals with posts is generally chosen.

Trusses are usually placed around the core to free the facade as in the example in the photograph.
In other cases, the truss can be placed within the facade, as shown in the two examples, in order to increase its lever arm.

For towers also arch-cables can be seen as trusses without diagonals, or as trusses in which the diagonals are not subjected to internal forces.

John Hancock Center in Chicago, 1969, Arch. Skidmore, Owings & Merrill, Eng. F. Khan (height of tower 344 m), X diagonals with posts

Bank of China in Hong Kong, 1989, Arch. I.M. Pei & Partners, Eng. Robertson, Fowler & Ass. (height 369 m), X diagonals without posts and V diagonals

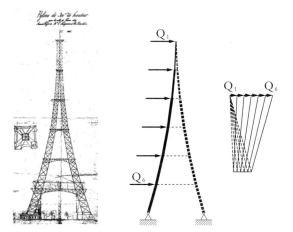

First Interstate World Center, Los Angeles, 1989, Arch. Pei, Cobb, Freed & Partners, Eng. P.V. Banavalkar (height 310 m), core with V and K diagonals

This principle was applied in the design of the shape of the Eiffel Tower. The purpose, in fact, was to limit the presence of diagonals in order to make the structure as transparent as possible. The engineer M. Koechlin, who first proposed the tower (patented in 1884), and later supervised its structural design as chief engineer in the Eiffel workshops, designed its shape based on these considerations.

Pylon of 300 m height for the town of Paris, France, 1884, Eng. M. Koechlin (project developed later with the help of Arch. Sauvestre and built under the supervision of G. Eiffel)

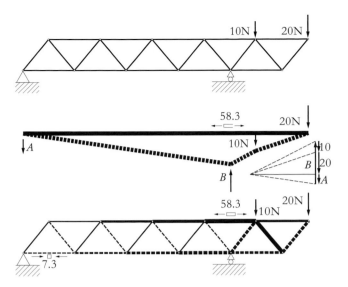

Triangular beam with cantilever: inversion of the internal forces in the chords due to the effect of loads on the cantilever

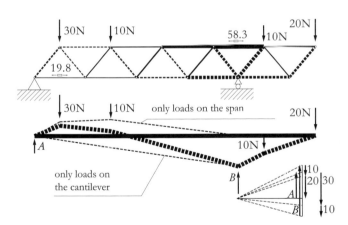

Truss with cantilever: effect of the loads on the span combined with those on the cantilever

A cantilever can also be attached to a truss, forming its extension beyond the support. The corresponding funicular structure can then be formed by an arch that starts at the end of the cantilever, descends toward the nearest support, and is then deviated by the force that acts there, and finally rises to the other support. At the latter, the horizontal component of the thrust is carried by the tie and conveyed to the other end of the arch. The support opposite the cantilever must carry a force that holds the structure to the ground and keeps it from falling over.

Again in this case, the internal forces in the truss can be studied, at least qualitatively, by means of the arch-cable. Because the tie is always above the arch, the upper chord of the truss will at all points be under tension, while the lower will be compressed for its entire length. Note that the situation is completely reversed with respect to that of the beam loaded along its span (between the supports).

The bars of the chords with the largest internal force are in the zone of the support, where the distance between the arch and the tie is largest. The diagonals under compression, as well as those under tension, are easy to identify by observing the slope of the arch.

Adding loads on the span, the situation is partially resta-bilized, with the lower chord under tension and the upper chord under compression. In the example shown here, the zone between the supports, with its loads, corresponds to the truss previously analyzed (p. 133), while the cantilever is identical to the one we have just examined. So it is possible to find the internal forces simply by adding together those already found for the two load cases.

In the left part the effect of the loads on the span prevails: the lower chord is under tension, while the upper is under compression. The internal forces, are always reduced by the action of the loads on the cantilever. Not only the cantilever, but also part of the span, on the other hand, are prevalently influenced by the loads acting on the cantilever.

These considerations can also easily be deduced from the arch-cable: to the right the arch is always below the tie, while to the left the situation is reversed.

Trusses with cantilevers

Gerber trusses

Therefore loads acting on the cantilever have a positive effect on the span, where the internal forces are reduced. This advantage can also be helpful in the case of multiple spans.

If we place simple trusses beside each other, the internal forces will clearly be identical to those of an isolated truss.

An alternative arrangement is to lengthen the first and third trusses with cantilevers and, in an alternating pattern, the following trusses, in order to be able to rest the other, shorter trusses directly on the ends of the cantilevers. The short trusses will not only have smaller internal forces, but will also transmit their load to the cantilevers of the long trusses, thus reducing the internal forces in those trusses as well. The diagram clearly shows the corresponding arch-cable. If the spans and the lengths of the cantilevers are regular, the distance between the arch and the tie of the span with the cantilevers (f_3) will be identical to that of the short trusses (f_2 and f_4), and much smaller than that of the simple trusses without cantilevers (f). It is also possible to choose the length of the cantilevers in such a way that the distances f_2, f_3 and f_4 are identical to the distance between the tie and the arches on the supports. In this way the maximum distances between arch and tie, like the internal forces in the chords of the trusses, will be halved with respect to those of simple trusses.

This system is called a "Gerber truss", from the name of the engineer who patented it in 1866, and designed its first application (in 1867) the bridge on the Main at Hassfurt. This bridge is also interesting for its form, which in the central part corresponds to that of the funicular structure of permanent loads, with a symmetrical arch and cable. The diagonals (N and X with posts) have the function of stabilizing the structure and carrying variable loads. In the lateral spans the form does not correspond precisely to that of the funicular structure of permanent loads. In this case the arch and the cable would have to cross again, but that would form an unstable structure and one not capable of carrying the variable loads (it is as if we had cut the bars in the chords of the first and third span, in the previous example, where the funicular arch crosses the tie).

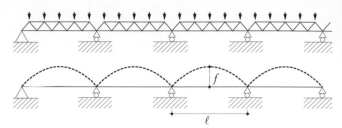

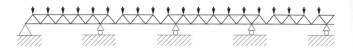

Series of simple trusses

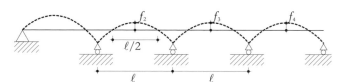

Series of simple beams supported by trusses with cantilevers: Gerber truss

Bridge on the Main at Hassfurt, Germany, 1867, Eng. H. Gerber (ℓ = 26.5 m, 42.7 m, 26.5 m)

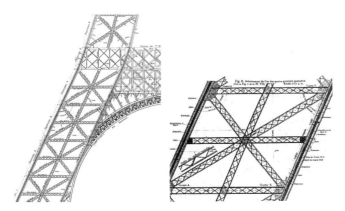

Eiffel Tower in Paris, 1889, Eng. G. Eiffel, M. Koechlin and
E. Nougier, Arch. Sauvestre, trusses forming the chords and the
diagonals

From a structural viewpoint the trusses studied thus far are all girders: simple girders, installed on two supports at the ends, cantilevers, girders with cantilevers or Gerber girders. But we have already seen other structural forms that, if we observe closely, also turn out to be trusses. This is the case in the following examples: the "Jawerth cable beam system" (see p. 55), which can be seen as a truss prestrained by tension so that it never receives compression in the chords, the "cables" that support the lateral spans of Tower Bridge in London (see p. 59; the arch, pylons and deck of the bridge built by Eiffel et Cie. in Oporto (p. 83; the arches and cables of the bridge on the Elbe at Hamburg (p. 118); the Eiffel Tower itself (see p. 150 and the illustration shown here), which while it can generally be considered a arch-cable, if analyzed in greater detail turns out to be composed of innumerable trusses. The chords, in fact, are subdivided into four bars connected by diagonals that, in turn, are composed of trusses.

**Trusses
for other
structural
forms**

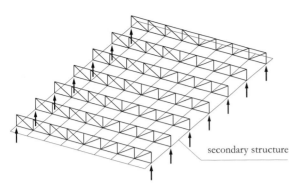

Series of trusses positioned on parallel planes

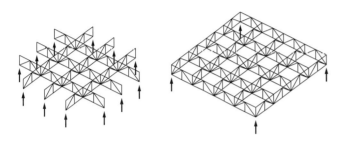

Lattice truss with supports on four sides, or supports at the four corners

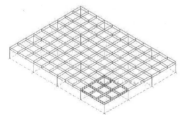

USM factory in Münsingen, Switzerland, 1963, Arch. B. and F. Haller

Let us now return to the most frequent use of trusses in architecture: the support of a flat roof. The simplest way to do this is to place a series of parallel trusses on the area to be supported. The permanent and variable loads on the roof will first be carried by a system of panels or secondary beams (purlins) and then transmitted by the main system of trusses to the supports placed on two sides of the roof.

The secondary system can be hung from the lower chord (see p. 140), resting on the upper chord (see p. 141) or attached alternately to the upper and lower chords to form what is known as a *shed*.

An alternative to the system of all parallel trusses is to arrange two series of trusses perpendicularly, to form what is called a *lattice*. The second series of trusses replaces the beams of the secondary system described in the previous example. These trusses do not, however, transmit the load only to the other beams; they also convey a part of it directly to their supports. In this case, for a rectangular roof, the supports are distributed on all four sides.

In these structures the loads do not necessarily have to follow the same direction right up to the supports. A load can also be transmitted diagonally, so the supports of a rectangular roof can be placed only at the four corners.

In the example shown here, the trusses are arranged in a square grid at intervals of 4.80 m, and are supported at the four corners of a square module that measures 14.40 m per side. These modules can be combined to obtain a large lattice supported every 14.40 m, both at the perimeter and inside its area.

Composition of trusses to support a roof

Lattice trusses

Space trusses

The efficiency of lattice trusses increases significantly if the diagonals are positioned to connect the upper and lower chords placed in the two directions. In this way we obtain a true spatial structure, where the bars form not only triangles, as in flat trusses, but also tetrahedrons and pyramids.

The first construction of this type was proposed by the physicist Alexander Graham Bell, better known for his contributions to the field of telephony. In 1904 he patented a system of prefabricated steel tetrahedrons that could be assembled to construct space trusses of different forms.

In the second half of the 20th century several systems were developed that allowed the assembly of space trusses using identical nodes and bars, pbaruced industrially.

The best-known system is the Mero, designed toward the end of the 1950s by Max Mengeringhausen. Up to eighteen tubular bars can be screwed into the node shown here, permitting great freedom in the construction of space trusses.

The lower illustration shows a space truss with large dimensions. This is a project for an aircraft hangar prepared by Konrad Wachsmann for the US Air Force in 1951. The objective was to build, quickly and in any situation, a large structure with overhangs of up to 50 m, composed of standardized parts that would not weigh more than 5 tons or be larger than 1 × 3 × 1 m, to facilitate transport.

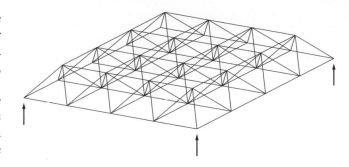

Space truss

Prefabricated tetrahedrons for the construction of space trusses, patented by A.G. Bell in 1904, used for flying machines

Mero prefabricated system, 1957, Eng. M. Mengeringhausen

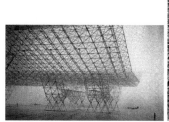
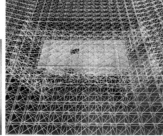

Project for a demountable aircraft hangar for the US Air Force, 1951, Arch. K. Wachsmann

Both lattice trusses and space trusses can also form the load-bearing structure of vaults and domes.

The illustration shows a pavilion vault in reinforced concrete built by P.L. Nervi, and the geodesic dome of Buckminster Fuller, which we have already seen (p. 104).

Vaults and domes composed of trusses

Hangar made with prefabricated reinforced concrete parts for the Italian air force, 1940, Eng. P.L. Nervi (lattice truss for a shell in the form of a pavilion vault with six supports, 100 × 40 m)

United States pavilion at the Montreal Expo, Canada, 1967, R. Buckminster Fuller (space truss for a geodesic dome, diameter 76 m)

SPACE TRUSSES

Beams

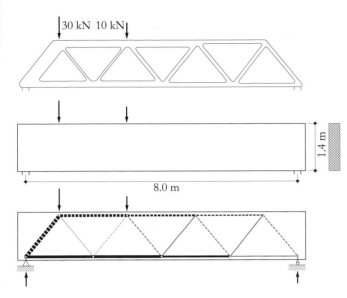

30 kN 10 kN

1.4 m

8.0 m

Truss and forces on a beam described using the truss analogy

Let us first imagine the truss shown here, in which the chords and diagonals have been dimensioned on the basis of the ultimate limit state (p. 28). The surfaces of the chords and diagonals have been calculated using the relationship

$$A_{\text{req}} = \frac{|N_{\text{d}}|}{f_{\text{d}}}$$

A properly dimensioned structure must have a minimum amount of material, i.e., the sections of the rods cannot be smaller than A_{req}. The structure can, however, have additional material as shown in the second figure. If the entire space between the rods is filled by material, we have a beam. In this case, the truss can be considered as a model that allows for the better understanding of the internal forces in the beam: rods in compression of the top chord carry the internal force in the compression zone of the beam, while the bottom chord corresponds to the tension zone. The internal forces in the diagonals are carried by the intermediate zone.

Internal forces in the middle zone of beams

In fact, the diagonals represent a simplification of the internal forces that are actually present. Because we have material between one diagonal and the next, it too will tend to carry part of the internal force. It is as if innumerable diagonals had been inserted, with the internal force distributed across all the material. If we isolate a tiny free body with the form of a small cube in the middle zone, we can observe two internal forces: one corresponds to the diagonal in compression, the other to the diagonal in tension. The same portion of material, then, is subjected to both tension and compression, but with lines of action of the forces that are perpendicular to each other. We have already seen a situation of this type in membranes, where the same part was subjected to tension in two directions, or in shells, where a compressive internal force was accompanied by one of tension or of compression.

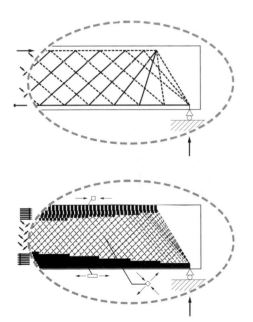

Middle zone occupied by overlaid diagonals and distribution of internal forces through all the available material (constant distribution of stresses in the zones under tension and compression)

Internal force in zones in tension and in compression

The upper zone in compression and the lower zone in tension (we have here a downward loaded simple beam) will also not be limited to the chord of the truss as hypothesized, but will tend to occupy all the available space. The zones near the edges will be subjected to the largest internal forces, resulting in a larger effective height. The precise distribution of the stresses in the tension and compression zones essentially depends on the behavior of the material (linear elastic or plastic), and is usually carefully studied by the engineer to permit an exact determination of deformations and strength. In the context of our analysis, we can simplify everything and hypothesize a constant distribution of compressive and tensile stresses. Actually this corresponds to the situation found in a beam made of a perfectly ductile material (steel for instance) and subjected to intense loads (the compression zone and the tension zone reach the plastic phase with complete yielding of the material).

Reinforced concrete beams

When a material mostly resists compression, while its tensile strength is limited, reinforcement is needed to carry the tensile stresses. This is the case, for example, of concrete, which is usually reinforced with steel. The reinforcement is usually placed where tensile stresses are foreseen: in the lower zone of our beam and in the middle zone in the direction of the diagonals in tension.

The illustration shows two possible steel reinforcements for a simple beam loaded at the center. We can see the longitudinal reinforcement that carry the tensile internal force in the lower part of the beam. In the upper example, the reinforcement of the middle zone is positioned vertically, as in the case of the posts under tension in an N-truss. In the second example, which is less common in practice, there are also inclined reinforcements, as if this were a beam with V-shaped diagonals. Furthermore, the lower longitudinal reinforcement is concentrated in the central part, where the internal force is largest. We can also see thin upper longitudinal reinforcement whose purpose is above all to attach the other reinforcement and to prevent it from moving during the pouring of the concrete. The concrete takes care of the compressive internal forces in this zone, so in static terms this reinforcement would not be needed otherwise.

The next illustration shows the cracks in a concrete beam subjected to a large load. In the lower zone the many cracks indicate that the concrete has completely relied on the reinforcement to carry the tensile force, while in the perfectly intact upper zone the concrete has fully performed its function, that of resisting compressive stresses. In the middle zone we see many inclined cracks: between them, a narrow intact zone of concrete functions as a diagonal under compression, while the reinforcement that crosses it carries the force, as the diagonals under tension would in a truss.

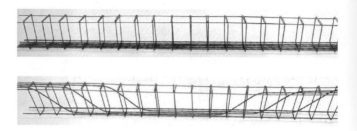

Steel bars to carry tensile internal forces in a simple reinforced concrete beam

Cracks in a reinforced concrete beam

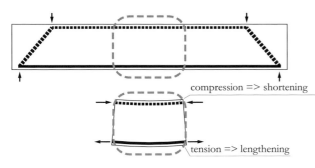

Bending of a beam and resulting deformations

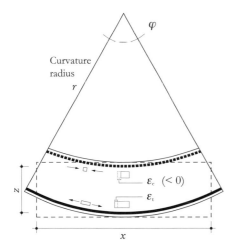

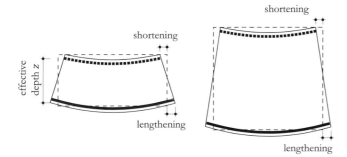

With the same deformation, the curvature is inversely proportional to the effective depth

When the funicular arch of the loads in a truss with horizontal chords is also horizontal (shear internal force is zero), the diagonals are not subjected to internal forces. Likewise, in the beam internal forces are limited to longitudinal compression and tension concentrated in the zones near the edges. This type of internal force distribution is known as simple bending.

Simple bending of a beam

The term «bending» is generally seen as a synonym for "folding" or "curving". In fact, the force in the tension zone will cause lengthening of the material, while the compression will cause shortening in the zone on the opposite side. By isolating a segment of a beam as a free body and indicating the deformations of the material, we can effectively observe the "folding" or "curving" of the beam.

Bending and curvature

So the curvature of a beam is the effect of bending (considered as internal force), just as the lengthening and shortening are the result of tensile or compressive internal forces.

Clearly, the greater the lengthening of the tension zone and the greater the shortening of the compression zone, the greater the curvature of the beam. Another parameter that influences the curvature is the depth of the beam or, more precisely, the distance between the center of gravity of the tension zone and that of the compression zone. From now on, we will call this internal distance the *effective depth z*. The illustration shows two beams of different depths with identical deformations. We can see that a reduction in effective depth corresponds to a larger curvature, and vice versa.

Having described how a beam subjected to bending is deformed, we can now examine its bending strength. This is reached when the entire available compression zone and the entire available tension zone are subjected to a stress equal to the strength of the material.

Strength of beams subjected to bending

As we have seen, the hypothesis of a constant stresses on the entire compression zone and the entire tension zone is actually valid only for ductile materials, like steel for example. For fragile materials, failure occurs when the maximum stress, usually located at the edge of the beam, reaches the strength of the material. We will not further investigate this theme but will continue our approach with constant intensity, which will permit us to outline the considerations that interest us, for the moment. To keep things simple, we will also consider a material with the same tensile and compressive strength f_d.

If we want to determine the loads that can cause the beam to fail (loads that correspond to its strength), we have to:
- insert a truss in the beam (or an arch-cable);
- determine the strength of the chords, multiplying the strength of the material (compressive and tensile strength f_d) by the area available for each zone (which in the case of a beam with a rectangular section is simply $A = t \cdot h/2$); and
- find the corresponding beam strength Q_{Rd} using a Cremona diagram.

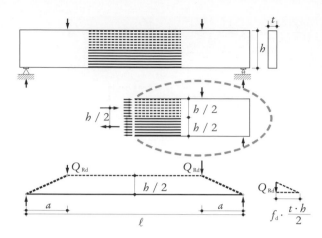

We can observe that the depth of the truss, or of the arch-cable, must correspond to the effective depth of the beam (distance between the resultant of the internal force in the tension zone and that of the compression zone). In the case of a beam with a rectangular section, the effective depth z at yielding corresponds exactly to half the total depth h. When the internal forces are smaller, the zones under compression and tension occupy a smaller space, so the effective depth z can be larger.

Strength of a beam with internal forces in the compression and in the tension zone (ductile material) and corresponding loads

Influence of the dimensions of a rectangular beam on its strength

We can now vary the dimensions of a rectangular beam to study their influence on its strength. First we will vary the thickness t, doubling it. The areas of the zones under tension and compression are thus also doubled, so their strength will also be doubled with respect to those of the beam with the initial thickness t. The form of the truss we insert in the beam, on the other hand, remains the same because the depth of the beam and the distance of the bars of the truss from the edge of the beam (which depends on the depth of the zones under compression and tension) have not changed. The result, then, is a Cremona diagram with doubled forces and unvaried slopes, so the strength new beam will also be doubled. This should not surprise us, if we consider the fact that the beam with double thickness ($2t$) can simply be interpreted as the sum of two beams with thickness t, placed next to each other.

Let us now double of the depth h without varying the other dimensions. Again in this case, the areas and the forces in the zones under tension and compression will be doubled but,

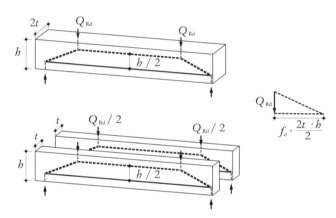

Influence of the thickness of a beam on its strength: doubling the thickness also doubles the strength (as if we had two beams)

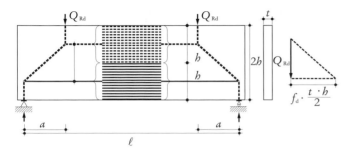

Influence of the depth of the beam on its strength: doubling h we obtain four times the strength

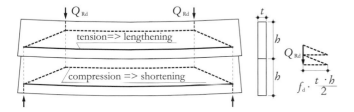

Comparison between a beam of double depth and two stacked beams

unlike the previous case, the effective depth z, along with that of the truss, will also be doubled. From the Cremona diagram we can see that with double horizontal forces and doubled slope of the inclined rod, the result with be a quadrupled strength.

At first glance this result may seem surprising: on doubling the depth, therefore doubling the material, we obtain quadrupled strength. Furthermore, if we simply put one identical beam on top of another, we only double the strength. But if we observe this situation carefully we can see that the load-bearing system and the corresponding deformations are essentially different from those of a beam with doubled depth: if the two stacked beams are not connected, two independent load-bearing systems are created, each with its own zones under tension and compression. The surface of contact between the two beams separates the zone under tension of the upper beam, which tends to lengthen, from the compressed zone of the lower beam, which tends instead to shorten. So we have sliding between the two beams. Only by stopping this relative movement, by gluing or welding the beams together, for example, can we obtain a transfer of forces between the tension zone, which can then occupy the whole lower beam, and the compression zone, which can then occupy the whole upper beam. Only in this case will we obtain quadrupled strength with respect to that of the original beam.

These considerations confirm what we have already seen with cables, arches and trusses: a structure is more efficient in terms of the use of material when its slenderness ratio (span/depth) is not too large. In other words, it is much better to position material in terms of depth than in terms of width. This conclusion is always valid provided that phenomena of instability are not present (lateral-torsional buckling for beams).

If we keep the section with its depth h and its thickness t and double the span ℓ and the other horizontal distances, we obtain the same dimensions and strengths of the zones under compression and tension, but a halved slope of the inclined rods, so the strength Q_{Rd} will also be cut in half.

From the last illustration (see p. 166, top) we can formulate a mathematical law that connects the dimensions of beams and the strength of material at the strength of the beam.

Considering the affinity between the triangle of the Cremona diagram and that formed in the structure by the inclined strut, we see that the strength of the beam Q_{Rd} is to the strength of the zone under tension (or compression) as the effective depth $h/2$ is to the distance a (distance between load and support). So we have the equation

$$Q_{Rd} = f_d \cdot \frac{t \cdot h^2}{4 \cdot a}$$

which sums up what we have just seen: the strength Q_{Rd} of the beam directly depends on the strength of the material f_d, the width t and the depth h squared (double depth, quadruple strength), while it is inversely proportional to the distance a. When we have to deal with a fragile material, the 4 in the denominator must be replaced with a 6, to take into account the non-constant internal force in the tension and compression zones.

Influence of the dimensions of a rectangular beam on its stiffness

By analyzing bars under tension or under compression, we have seen that the stiffness of these simple structures depends on the stiffness of the material (modulus of elasticity E), the area of the section and the length of the bar (see p. 18). We can also find similar relationships for beams in bending. Again in this case, we define the stiffness of the structure as the ratio between the load Q we apply and the deflection w that we cause.

The influence of the thickness t is easy to describe: on doubling it, we obtain half the internal forces in the beam, so the deformations and deflection will also be cut in half. Therefore we have doubled the stiffness, exactly as if we had placed two identical beams next to each other; in other words, the stiffness is directly proportional to the thickness.

If we double the depth h, keeping the load Q and the other measurements constant:

- the effective depth z doubles, so the forces in the chords of the funicular structure that represents the functioning of the beam will be halved;
- the areas of the zones under compression and tension will also double ($t \cdot h$ instead of $t \cdot h/2$);
- the tensile and compressive stresses we obtain by dividing the force by the area will, then, be a quarter of that of the original beam;
- so the strains of the zones under tension and compression we obtain by dividing the stresses by the modulus of elasticity will also be reduced to a quarter;

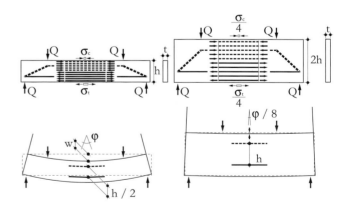

Internal forces, deformations and deflections in a beam with depth h and a beam of twice that depth

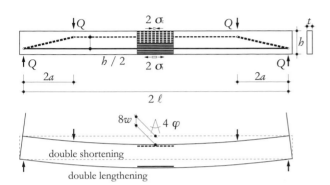

Internal forces, deformations, curvature, bending angle and deflections in a beam of double length

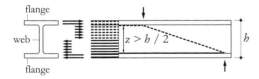

Wide-flange section

– As previously shown (see p. 165), the curvature is proportional to the deformations, while it is inversely proportional to the effective height z: in our case with a double depth, then, we get displacement reduced to 1/8.

(see p. 165)

In other words, by doubling the depth h we obtain a stiffness that is eight times larger (the displacement corresponds to that of eight beams placed side by side, or stacked, without being connected!). In fact, the stiffness of a rectangular beam depends on its depth cubed.

Let us now double the span ℓ and the distance a, keeping the section of the beam and the load constant: the internal forces and deformations will also double, so the curvature will double as well. The angle of bending that depends on the curvature and the length, then, will be quadrupled, while the deflection, which depends on the bending angle and the distance, will be eight times greater than that of the beam with span ℓ. To generalize, we can say that the stiffness of a beam bearing concentrated loads is inversely proportional to its length cubed.

To sum it up, with a constant ratio between the distance a and the span ℓ, we can say that the stiffness of the beam, defined as the ratio between the load Q and the deflection w, is proportional to

$$\frac{E \cdot t \cdot h^3}{\ell^3}$$

If we now consider a uniformly distributed load q, we will also have the resultant of the loads depending on the span ($Q = q \cdot \ell$) for which the deflection w will depend on the span to the fourth power, and the stiffness will depend on

$$\frac{E \cdot t \cdot h^3}{\ell^4}$$

The most efficient sections: wide-flange sections

The rectangular sections we have examined up to this point can also be improved to obtain greater efficiency (strength and stiffness) while keeping the quantity of the material used the same. As we have already seen, we can increase efficiency by reducing the thickness t and increasing the depth h. But we want to look for other possible optimizations, keeping the total depth h constant. Imagine removing part of the material from the lower side of the compression zone and putting it in the upper part to widen it. Repeating the same operation with the tension zone, we obtain a section with the same area and the same depth, but with a much larger distance between the center of gravity of the tension zone and the compression zone. Sections of this type, called wide flange sections (or double-T sections), are often used in steel constructions. They can be easily produced with the process of lamination (steel), or extrusion (aluminium). The horizontal parts, whose role is to carry the tensile and compressive forces due to bending, are called flanges. The vertical part, known as the web, carries the forces of tension and compression that correspond to those in the diagonals of a truss.

Influence of the dimensions of a wide-flange beam on its bending strength and stiffness

Therefore a beam mainly subjected to bending, it is best to choose a section with wide, thick flanges and a slender web. Actually beyond a certain limit, when the flanges are too wide and the web too slender, the material cannot be fully exploited, since transverse deformations can cause local instability before the strength of the material is reached (buckling phenomenon).

Let us overlook that effect, for our purposes, and consider an ideal section in which all the material is arranged in the flanges. In this case the effective depth of the section, seen as the distance between the centers of gravity of the zones under tension and compression, approaches the total depth h (more exactly, $z = h - t_f$). So we have an effective depth that is almost twice that of the rectangular section.

Again in this case, the strength of the beam can be deduced from the Cremona diagram, considering its affinity with the triangle formed by the inclined rod, the effective depth z and the distance a between the load and the support. The beam strength Q_{Rd} is to the strength of the zone under tension (or compression) $f_d \cdot t_f \cdot b$ as the effective height $h - t_f$ is to the distance a. From this ratio, we obtain:

$$Q_{Rd} = f_d \cdot t_f \cdot b \cdot \frac{(h - t_f)}{a} \approx \frac{f_d \cdot t_f \cdot b \cdot h}{a}$$

which tells us that the strength depends, linearly, on the depth and not on its square as in the case of a rectangular section. If we now consider the stiffeness of the beam, it depends linearly on that of the material (elasticity modulus E) and the area of the flanges $b \cdot t_f$.

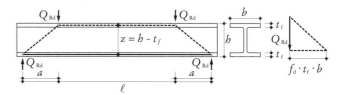

Influence of the dimensions of a wide-flange beam on its bending strength and stiffness

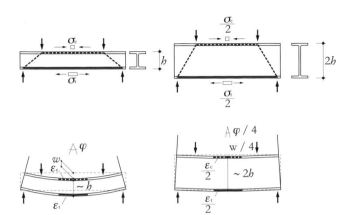

Influence of depth on stiffness

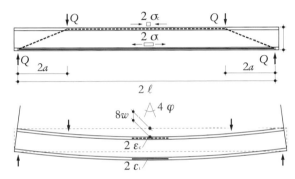

Deformation of a beam with span ℓ and distance a doubled

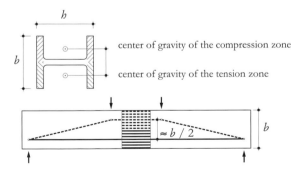

Beam with wide-flange section arranged with the flanges vertical: if the dimensions h and b are equal, the effective depth is almost halved with respect to the same beam with horizontal flanges, so we obtain half of the strength and a quarter of the stiffness

To find the influence of depth h, let us consider a beam of double depth. Under the same load, we will have half the internal force in the flanges and, therefore, their deformation will also be halved. The bending angle like the deflection w, will therefore be reduced to a quarter. In other words, the stiffness is quadrupled. The stiffness, then, depends on the depth h squared.

Let us now double the span and the distance a, as we did with the rectangular-section beam. We will have internal forces and deformations that are doubled in the flanges; the curvature also doubles, but the bending angle is quadrupled and the deflection is eight times greater: again, the stiffness is inversely proportional to the span cubed in the case of a concentrated load, and to the fourth power in the case of a distributed load.

To summarize, when the thickness of the flange t_f is small with respect to the depth h, and the effective depth is approximately equal to the depth, we have the stiffness of the beam expressed as the ratio between the load Q and the deflection w, proportional to the expression

$$\frac{E \cdot b \cdot t_f \cdot h^2}{\ell^3}$$

If we rotate our beam by 90° and keep the downward loads the same, the compression zone will still be in the upper half and the tension zone will still occupy the remaining material in the lower part. The areas of these zones, although they are subdivided between the two flanges, are equal to those of the beam before rotation. But the center of gravity of the compression zone is lowered and that of the tension zone is raised, so the effective depth becomes much smaller. In a section where h and b are equal, the new effective depth (equal to $b/2$) is slightly more than half of that of the previous example ($h - t_f$). So the strength of the beam is reduced by half.

A section positioned in this way is also less efficient in terms of stiffness. Keeping the load constant, we obtain double internal force on the material due to the halving of the effective depth. The bending angle which as we have seen depends both on the strain and the height, will be further doubled, so the deflection is quadrupled and the stiffness is reduced to a quarter of that of the same beam with horizontal flanges.

Behavior of a wide-flange beam with vertical flanges

Therefore we should avoid positioning wide-flange beams with the flanges vertical when the controlling loads are also vertical. Precisely to avoid misunderstandings in this matter, we use the term wide-flange section or double-T section, and avoid the term H-section.

The efficiency of a section

The efficiency of a section can therefore be determined by the ratio between the effective depth z and the total depth h. The more the material is positioned near the upper edge and the lower edge, the greater the efficiency of the section. The illustration shows some sections in increasing order of efficiency, from left to right.

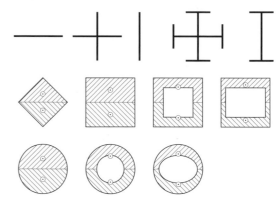

Comparison of the efficiency of different sections (for vertical loads), shown in rising order from left to right (the asterisks indicate the center of gravity of the zones under compression and tension)

Form, section and structural efficiency

As we have seen, the structural efficiency of a beam depends on its section and its depth (height × 2 of a rectangular section => strength × 4 and stiffness × 8), or on the span/depth ratio, known as the slenderness ratio. The illustration shows the quantity of material needed to support a concentrated load Q_d, with a beam across a span ℓ in relation to the slenderness ratio. The diagram also shows the curves that refer to cables, arches, cable-arches and trusses.

The efficiency of a beam with a constant wide-flange section is similar to that of a truss with constant chords. With an identical effective depth, their internal forces are also comparable. Moreover, the quantity of material necessary for the diagonals of the truss is comparable to that needed for the web of the beam, which has the same function and is subjected to similar internal forces.

The lower efficiency of rectangular-section beams with respect to wide-flange beams is also clearly visible in this diagram: for equal slenderness, the compact sections require much more material.

Exactly as in trusses, where by varying the area of the rods in response to the internal forces it is possible to reduce the quantity of material required, so in beams a variation of the section along the length leads to considerable savings in terms of material. This reduction, however, is often purely theoretical due to practical difficulties. For example, it is very hard to vary laminated sections and impossible to vary extruded ones, unless the surplus material is removed after production.

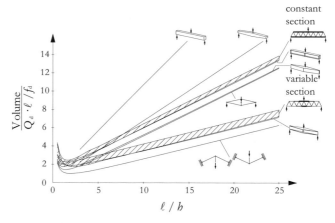

Quantity of material in relation to the slenderness ratio ℓ/h (for the same span ℓ, load Q_d and material strength f_d)

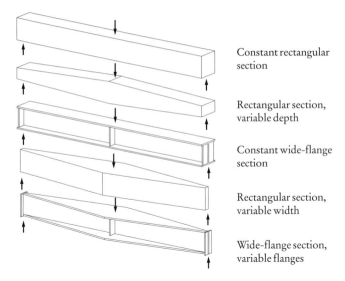

Constant rectangular section

Rectangular section, variable depth

Constant wide-flange section

Rectangular section, variable width

Wide-flange section, variable flanges

Possible variations of sections of a beam loaded at the center

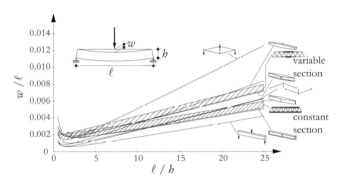

Deflection at mid-span caused by a concentrated load in relation to the slenderness ratio ℓ/h (maximum unitary deformation in the material: $\varepsilon = 0.001$)

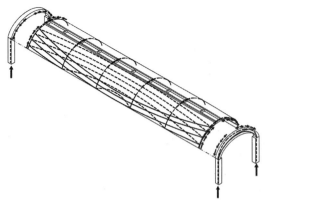

Cylindrical shell functioning as a beam, of the Kimbell Art Museum (see p. 109)

The illustration shows several possibilities in order of decreasing efficiency, from bottom to top (as in the diagram). In wide-flange beams the variation of the section leads to adaptation of the width of the flange, or its thickness, to the internal forces. With a load acting at mid-span these dimensions, zero at the supports, vary linearly, reaching a maximum at the center. Stiffeners (plates placed vertically, at the supports and the central load) are required in these beams to carry the applied force and transmit it to the web.

In rectangular sections, too, it is possible to vary the width or the depth. In the first and more efficient case, the width on the supports is necessary to permit the beams to carry the internal shear forces, which in trusses are found in the diagonals.

A structure is more efficient not only when a smaller quantity of material is required to guarantee strength, but also when it has greater stiffness. The diagram shows the deflections at mid-span of our beam, once again in relation to the slenderness ratio. It shows the curves that refer to the structures previously discussed, to permit direct comparison. Beams with a constant section, where a large part of the material is superfluous, have smaller deflections. This material, although in excess where strength is concerned, has a positive effect on deflections. But we should note that by arranging it in the zones with the most stress, its efficiency is increased further, because with the same load the maximum internal force in the material and the unitary deformation are diminished.

Cylindrical shells like those of the Kimbell Art Museum (see p. 109) behave in a manner similar to beams in the longitudinal direction, with a tension zone along the lower edges and a compression zone along the apex of the shell. Indeed, in a simplified manner, they can be considered as beams with a curred section. In the diagram the beam function is shown schematically with several stacked arch-cables. Transversally, the shell functions like a vault, however, that discharges its thrust onto the longitudinal system. At the two ends the force cannot be directly transmitted to the supports. In fact, the arch-cables are not on vertical planes, so they transmit the internal forces to the supports with a horizontal component. Therefore an end diaphragm is required, capable of carrying this component of the forces and conveying the vertical component to the supports. Note that these diaphragms placed at the ends, although they have an arched form, function as beams, given the lack of an element capable of carrying the transverse thrust on the supports.

Simple beams with concentrated and distributed loads

As in the case of trusses, we will use the term simple beam to indicate a beam resting on supports at the two ends. The considerations regarding strength, stiffness and efficiency outlined thus far for beams subjected to concentrated loads are also clearly valid in the case of distributed loads.

While on the one hand the functioning can be understood in all its components by imagining a truss inserted in the structure, on the other hand both the most important internal forces and the general functioning can also be described with the help of the arch-cable.

The illustration shows the constructions of an inserted truss and an arch-cable for a simple beam subjected to a concentrated load at mid-span, and to a uniformly distributed load. Once the effective depth z has been determined, which depends essentially on the form of the section, it is possible to quantify the internal forces of the zones under tension and compression.

Considering the affinity between the triangle of the Cremona diagram and the one formed by the effective depth z and the distance $\ell/2$, as we have already established for the cable (see p. 40), for a beam with a concentrated load we have:

$$N = \frac{Q \cdot \ell}{4 \cdot z}$$

where N, besides being the internal force in the tie of the arch-cable, is also the maximum force in the compression zone and the tension zone at mid-span of the beam.

The case with a distributed load can be correlated with the previous case, considering the concentrated force Q as the resultant of all of the distributed force $q \cdot \ell$ and taking into account the fact that the rise of the triangular arch of the resultant is double with respect to that of the parabolic funicular arch. By replacing Q with $q \cdot \ell$ and z with $2 \cdot z$ in the previous equation, we obtain the new relation:

$$N = \frac{q \cdot \ell^2}{8 \cdot z}$$

Comparing these two equations, we see that the internal force is halved when the load is distributed. This comes from the fact that the load exerted near the supports has a much smaller effect on the central part of the beam. These equations can also be derived by dividing the bending movement at mid-span by the effective depth z (see appendix 3, p. 236).

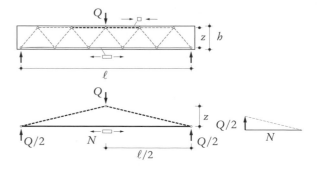

Simple beam with load concentrated at mid-span

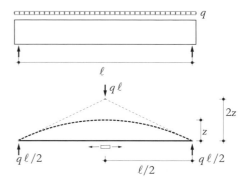

Simple beam with distributed load

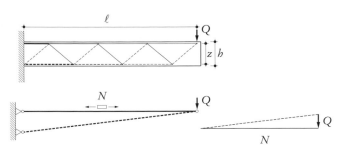

Cantilever with concentrated load Q

Cantilever with distributed load q

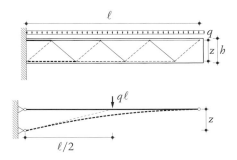

Trees with the form of the trunk that follows the pattern of the internal forces (Pinus pinaster, Prunus cerasus, Quercus robur and Tilia cordata)

CN Tower in Toronto, Canada, 1973-1976, Arch. E.R. Baldwin, Eng. F. Knoll, A.G. Davenport, B. Thürlimann, h=553 m

Cantilevers

Cantilevers can also be studied by inserting a truss cantilever and analyzing a corresponding arch-cable composed, for example, by a lower arch and an upper tie. From the conditions of equilibrium of the latter we can directly determine its internal force, which corresponds to the maximum tensile and compressive force in the truss and in the beam.

In the case of a load concentrated at the end we have

$$N = \frac{Q \cdot \ell}{z}$$

while with a distributed load, considering the fact that its resultant $q \cdot \ell$ acts at the middle of the cantilever, we obtain

$$N = \frac{q \cdot \ell^2}{2 \cdot z}$$

The strengths of the beams can be determined by comparing the internal forces with the strengths of the tension and compression zones. As we have seen, we have $N_{Rd} = f_d \cdot t \cdot h/2$ and $z = h/2$, in the case of the rectangular section, and $N_{Rd} = f_d \cdot b \cdot t_f$ and $z = h - t_f$, for the wide-flange section with a slender web.

A cantilever in which the sections follow the pattern of the internal forces has its material concentrated above all in the clamping zone, while at the end the section can be tapered.

One good example of this arrangement is the form of a tree trunk, whose main structural function is to carry and transmit to the ground the horizontal wind pressure. Very tall buildings and towers are also subject to stressed mostly by wind. This is why towers are often tapered, with a variation of their section similar to that of tree trunks.

Study of the truss and its internal forces allows us to also predict the deformations. If we consider that the upper zone, under tension, will tend to lengthen, while the lower, compression zone tends to shorten, we will have a cantilever that bends downward. Unlike a simple beam, where the center of curvature was above the beam, in the case of the cantilever it is found below the structure.

Furthermore, because the internal force varies from a maximum at the clamping point to a minimum (zero) at the end, the curvature will also follow the same pattern.

In general, to understand how a beam bends it is therefore sufficient to consider the internal forces: the center of curvature will also be located on the part where the beam is under compression.

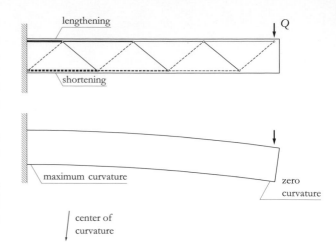

Internal forces, deformations and deflections of a cantilever

Beams with cantilevers

Exactly like trusses (p. 151), by adding one or more cantilevers to a simple beam a reduction of internal forces in the span can be obtained. When the load is uniformly distributed along the length of the beam, the longer the cantilevers, the greater the reduction of the internal forces in the span, while those close to the supports will increase.

The illustration shows the corresponding arch-cables for some beams subjected to a uniformly distributed load. The span ℓ of the beams, taken as the distance between the supports, is always equal, so the rise f relative to the central part of the arch-cable also remains the same. In the five examples shown here, the length ℓ' of the cantilever varies. The effect is to raise the tie and to subdivide the rise in the central span into an upper part and a lower part. This also leads to variation of the internal forces in the beam. We should remember that the distance between the arch and the tie of the arch-cable is proportional to the internal forces in the truss with parallel chords or in the beam. The effect on the internal forces in the beam of a cantilever that is not too long, therefore, is beneficial: it reduces the internal forces in the span where the forces are generally decisive, and increases them close to the supports.

When ℓ' reaches half of the span ℓ, the arch is always below the tie. This particular case actually functions as four cantilevers that balance each other in pairs.

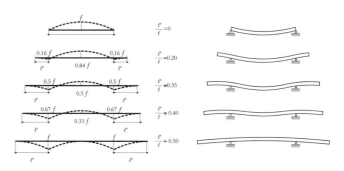

Beams with cantilevers: corresponding arch-cables and deflections

In this case the beneficial effect of the cantilever is completely cancelled out, because the internal forces close to the supports, which are now decisive, have reached the intensity of those that can be found at mid-span in a beam without cantilevers.

The situation is much more balanced when ℓ' equals about $0.35\,\ell$ (more precisely $\ell/\sqrt{8}$) so that the distance between the arch and the tie in the span is identical to that at the supports. In this case the maximum internal forces are halved with respect to those in a simple beam without cantilevers. This situation is interesting when the section of the beam is constant: the structure is exploited equally both at the supports and in the span, so if the dimensions are those required by the ultimate limit state criterion, the quantity of excess material is reduced to a minimum.

Next to the arch-cables structures, the illustration also shows beams with their respective curvatures and resulting deflections. It is interesting to observe how cantilevers, when they are not too long, also have a beneficial effect on the deflections of the beam.

In 1917 D'Arcy W. Thompson pointed out, in his work *On Growth and Form*, that many organisms in the animal world have forms that are similar to the distribution of internal forces in structures created by man. In particular, quadrupeds, both mammals and reptiles, can be seen in structural terms as beams with two cantilevers, represented by the head and the tail. Thompson made diagrams of the bending moments of certain quadrupeds and described their similarities to the real form of the animal (see the illustration). The diagram of the bending moments has exactly the same development as that of the arch-cables (see p. 134). These similarities are perfectly compatible with the theory of evolution: when the internal forces on the "beam" are large, an organism evolves by increasing the height of the "structure", so the bones, with their function of compressive strength, and the tendons, under tension, do not undergo forces that are too intense. Thus a species can increase its efficiency and have a greater possibility of survival.

In the same manner in architecture, a beam can be given a similar form in order to place the material where it is effectively under the largest internal force.

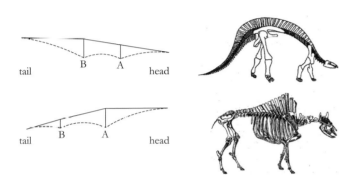

Diagrams of bending moments represented by D'Arcy W. Thompson (similar to arch-cables) for a dinosaur with a very long, heavy tail, and a quadruped with a long neck and a heavy head

Project for the intercontinental airport of Fiumicino in Rome, Italy, 1957, Arch. + Eng. P.L. Nervi

Gerber beams

Just as in trusses, it is possible to combine simple beams with cantilever beams to obtain a single structure that extends over multiple supports.

Note that the criterion we have introduced to test the static determinacy of a truss (see. p. 128) also works for beams. On applying it to the example shown here with six reactions (five vertical and one horizontal), four bars (or beams) and five nodes we obtain $6+4=2 \cdot 5$. We can lengthen this system at will, adding a pair of beams and two supports at a time. Because each pair of beams we add corresponds to two additional nodes, the equation becomes $8+6=2 \cdot 7$, $10+8=2 \cdot 9$, and so on.

When analyzing the cantilevers of a Gerber beam we have to consider the action of the distributed load and of the internal force transmitted by the support of the short beams. The cantilevers are therefore under greater internal force than those previously examined. If, in a Gerber beam, we want to make the internal forces in the short beams have the same intensity as that on the supports of the long beams, we have to reduce the ratio ℓ'/ℓ previously defined. For a uniformly distributed load, an optimum situation is achieved when ℓ'/ℓ is equal to about 0.15.

It is useful to respect this proportion when the section of the beam is constant throughout its length: in the governing zones, in the span and on the supports, equal internal force on the material can be obtained in this way.

Of course it is also possible to vary the depth of Gerber beams, to adapt them to internal forces. In the example shown here, the short beams resting on the cantilevers reach their largest depth in the span. The long beams, on the other hand, have a larger depth at the supports, which is where the internal forces are largest in this case.

In theory, due to the effect of permanent loads, the smaller internal forces in the long beams are located where the arch of the arch-cable structure meets the tie. Again in this case, exactly as in the lateral spans of the bridge of H. Gerber (see p. 152), a sufficient depth is maintained in those zones to stabilize the structure when it is subjected to variable loads.

Static determinacy of a Gerber beam

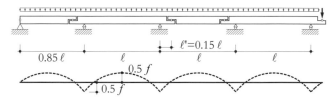

$0.85\,\ell$ ℓ $\ell'=0.15\,\ell$ ℓ ℓ

$0.5\,f$

$0.5\,f$

Gerber beam with identical internal forces in the span and at the supports

Highway viaduct over the Setta River near Bologna, Italy, 1958, Eng. R. Morandi

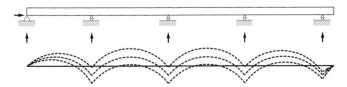

Static indeterminacy of a continuous beam and possible corresponding arch-cables structures

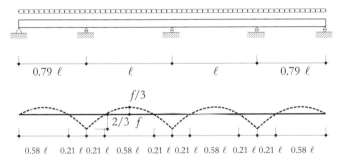

0.79 ℓ   ℓ   ℓ   0.79 ℓ

f/3

2/3 f

0.58 ℓ   0.21 ℓ 0.21 ℓ   0.58 ℓ   0.21 ℓ 0.21 ℓ   0.58 ℓ   0.21 ℓ 0.21 ℓ   0.58 ℓ

Internal forces in a continuous beam with constant stiffness and linear-plastic behavior, with only a uniformly distributed load

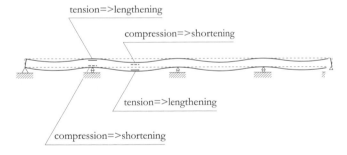

tension=>lengthening

compression=>shortening

tension=>lengthening

compression=>shortening

Deflections of a continuous beam

Continuous beams

If we place a continuous beam on the supports of the Gerber beam studied previously, we obtain a statically indeterminate system: $6 + 1 > 2 \cdot 2$. It is as if we had eliminated the hinges, so in those points the arch-cable will have an indefinite position and we will no longer know its point of intersection with the tie. In fact, we can draw an unlimited series of arch-cables, all in equilibrium with the external loads.

Actually there will only be one state of internal forces that can be described with a well-defined funicular structure. This will depend, however, not only on the loads but also on the mechanical behavior of the material, the distribution of the stiffness along the structure, and a whole series of actions such as, for example, temperature variations and displacement of the supports.

If we assume that there is no other action besides the load (i.e. once the load has been removed the internal forces are zero at all points of the beam) and that the behavior of the material is linear-elastic and that the section of the beam together with its stiffness is constant, the arch-cable structure corresponding to a continuous beam with regular spans, subjected to a uniformly distributed load across its length, has a distance from the tie that, in span, corresponds to one-third of the total rise, and on the supports reaches the remaining two-thirds. In other words, the internal forces in the compression zone and the tension zone on the supports are twice those found at mid-span.

We can also observe that the situation is similar to that of cantilever beams and Gerber beams: in the span, the upper zone is under compression while the lower zone is under tension; at the supports, the situation is the opposite: the upper zone is in tension and the lower is in compression. In fact, the internal forces are identical to those of a beam with cantilevers with $\ell' \approx 0.4 \cdot \ell$, so the slope of the beam on the supports becomes zero. This is due to the fact that in the continuous beam with equal spans, the zone above the supports must remain horizontal, because that section corresponds to an axis of symmetry.

The side spans, if they are as long as those at the center, will have the larger internal forces because the funicular arch, meeting the tie on the last support, will have a greater height in the span. To prevent this effect we have to reduce their span. If the latter equals about 0.8 times the length of the central spans, equal to the distance between the support and the meeting point of the arch and the tie, the internal forces will be identical in all the spans. This criterion is usually applied in the subdivision of the spans of a continuous beam bridge as in the example shown here.

The internal forces in the zone of the supports of a continuous beam bridge are generally larger than those found in the span. A continuous beam where the sections are adapted to the internal forces will therefore assume a familiar form, with the greater height at the supports and tapering along the span. This also leads to a beneficial concentration of permanent loads near the supports.

Example of a continuous beam bridge with side spans spans 0.8 as large as the central spans (bridge over the Tiber in Rome, Italy, 1972, Eng. S. Zorzi)

Bi-clamped beams

If from a continuous beam we isolate a free body that includes one central span, we have the situation shown in the illustration. At the point in which the beam has been cut we have to introduce the internal forces: tension in the upper part, compression in the lower part, and an internal force in the middle zone that corresponds to the force transmitted from the supports. These internal forces are very similar to those found in the clamping of a cantilever.

In the case of a uniformly distributed load, at the clamp, where the distance between the arch and the tie reaches 2/3 of f, we have an internal force, in the compressed zone and the one under tension, equal to:

$$N_{\text{clamp}} = \frac{q \cdot \ell^2}{12 \cdot z}$$

While at mid-span, where the distance between the arch and the tie is half, the internal force is also halved:

$$N_{\text{span}} = \frac{q \cdot \ell^2}{24 \cdot z}$$

The deflections are also compatible with this new situation: exactly as in the continuous beam with regular spans and distributed load, in the clamped beam if the support structure is very stiff, the angle of rotation at the ends will be zero.

The illustration shows an example where the depth of the beam follows the intensity of the internal forces. As we can

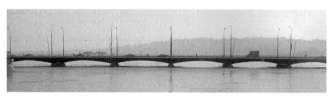

Continuous beam with depth adapted to internal forces (Quai Bridge in Zürich, 1882-1884, Eng. A. Bürkli).

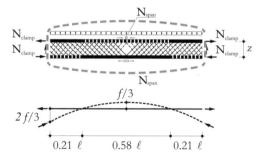

Bi-clamped beam, internal forces from a uniformly distributed load (linear-elastic material, constant stiffness)

Bridge over the Marne at Luzancy, 1946, Eng. E. Freyssinet
(ℓ = 55 m; height at mid-span 1.22 m)

see, the clamped beam is much higher than the span section. In fact, in a beam of this type the internal forces at the clamp can even be more than twice as large as those in the span, as previously calculated. This is the result of the permanent load that is concentrated near the clamps and, above all, of the greater stiffness of the beam in the clamp zone, which causes an increase of the internal forces at the clamp and a reduction in the span.

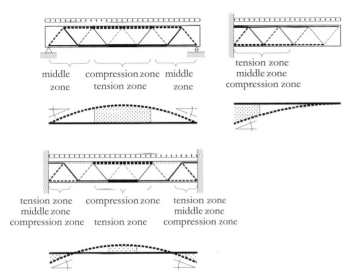

Zones with the largest internal forces in simple beams, cantilevers and clamped beams

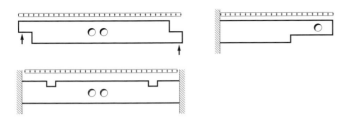

Openings and weakenings in zones under smallest internal forces

Zones of greater or lesser internal forces in beams

When analyzing trusses, we explained how to find the bars with the largest internal force by studying the corresponding arch-cable. The internal force in the chords is proportional to the distance between the arch and the tie, while the internal force in the diagonals will be more intense at the point at which the arch has the largest slope. This procedure also works for beams.

The illustration shows the zones with the largest internal forces for the most common types of beams: simple beams, cantilevers and clamped beams. These three elements can also be seen as the components of beams with cantilevers, of Gerber beams, continuous beams and frames.

Knowledge of the zones with the largest internal forces can be very useful, both because this data is needed for dimensioning, and because these zones should be protected against cost cutting or any other factors that might weaken them. For the same reason, it is also useful to know which zones have the smallest internal forces: openings or reductions of the section in those zones will not cause major problems for the structure.

The examples shown here indicate the most advantageous positions for openings and weakenings. We should remember that wherever a zone, even with limited internal forces, has been cut away, it is always necessary in any case to find an alternative flow path for the internal forces.

Frames

In the Luzancy bridge, as we have just seen, the beam is clamped in two very rigid vertical parts (p. 181) sustained by two fixed supports. In fact, if we look at the whole arrangement, this structure should be considered as a *frame*. This is the term for a structure composed of horizontal parts (beams) and vertical parts (columns or side walls) connected in a monolithic way to obtain a single structure.

Their functioning is identical to that of an arch in which the line of action of the internal forces emerges beyond the material. In other words, an arch whose form is significantly different from that of the funicular polygon of permanent loads can be definied as a frame (see p. 82).

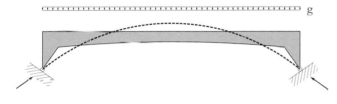

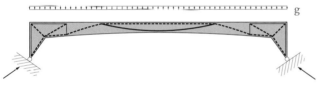

Longitudinal section, funicular arch and internal forces for the bridge on the Marne at Luzancy, France, (two-hinged frame), 1946, Eng. E. Freyssinet (see p. 181)

Two-hinged frames

In the case of the bridge over the Marne at Luzancy (see figure and diagram) we have a two-hinged frame so, as in the case of arches, we have a statically indeterminate system. The funicular polygon of loads must pass through the two hinges at the base of the structure, therefore the span is definite while the rise is indeterminate. The diagram shows the funicular polygon of the permanent loads that is formed when there are no other actions. If the temperature rises the polygon will tend to lower, while it will rise if the temperature drops, or if the terrain yields slightly and the bases of the supports move further apart.

We have already seen how to describe the functioning of this type of structure, by determining the tension and compression zones. Where the funicular polygon remains inside the structure, we have only compression; where the funicular arch extends beyond the material, we have a compression zone along the surface closest to the funicular polygon and a tension zone on the opposite side. We have already described the internal forces inside a beam clamped at the ends, and we have already encountered the internal forces of compression and tension found in the side walls. In this bridge, the side walls have been designed to place material only where it is needed. Thus there are external ties and compression elements on the inside. In the clamping zone, the zone under tension and the zone under compression operating at the end of the beam connect with the elements of the side walls that are subjected to the same type of internal forces. Because they are not on the same line of action, they generate two forces of deviation that are carried by a diagonal element in compression.

Three-hinged frames

When we have a structure with three hinges, as in the example shown here, the rise of the funicular polygon of loads is known thanks to the static determinacy of the system. The funicular curve must pass through the three hinges. Note that the funicular polygon is influenced only by the loads, while temperature variations and shifting of the supports have no influence on the internal forces. As in the previous example, this too is a prestressed reinforced concrete bridge. This means that the internal forces of compression are carried by the concrete parts, while pre-stressing cables of very high strength steel have been inserted to handle the forces of tension. In fact, as shown in the diagram, the positioning of the prestressing cables precisely follows the tensile internal forces: on the exterior of the side walls and on the extrados of the beam.

Other cables are also placed vertically in the webs of the bridge deck. In fact, the system with arches and ties we have sketched corresponds to the one we have defined as a arch-cable, which as we know corresponds to a simplification of the functioning. A step closer to the actual load-carrying system is represented by the truss: the vertical cables we can observe correspond to its posts and are needed to carry the corresponding tensile forces. In the truss we also have the upper chord, which in this case is under tension, and the lower chord, which is under compression. In the actual section they are located in the deck slab of the bridge and the lower slab of the box section. Unlike the arch-cable, in the truss the internal forces in the chords vary perceptibly along their length. This can also be seen in the bridge we are analyzing: the cables of the deck slab are mainly concentrated in the clamping zone and the thickness of the lower concrete slab varies from a few centimeters near the hinge at the key to 1.00 m in the clamping zone, where the internal forces are largest.

Bridge on the Tagliamento River at Pinzano, 1969, Eng. S. Zorzi (three-hinged frame, $\ell = 163$ m, $b_{span} = 2.50$ m, $b_{clamp} = 7.00$ m), layout of the prestressing cables and description of the functioning by means of an arch and a truss

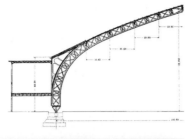

The illustration shows another three-hinged frame but in a completely different situation. The steel structure of the *Galerie des Machines* for the Universal Exhibition of Paris in 1889 is composed of a series of frames in which all the elements have the form of trusses. The form is based above all on functional considerations. The gigantic nave had to host the emblems of the industrial revolution, large machinery and products of heavy industry.

Galerie des Machines at the Universal Exhibition of Paris in 1889, Arch. F. Dutert, Eng. V. Contamin ($\ell=115$ m, $f=45$ m, length of nave 420 m)

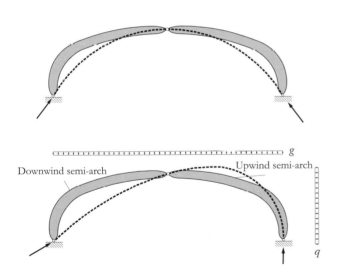

Downwind semi-arch   Upwind semi-arch

Galerie des Machines: funicular polygons corresponding to the permanent loads, and overlapping wind and permanent loads

In evaluating the general behavior of the structure, we have to also consider the variable loads, namely wind, which cannot be neglected in a relatively light structure of this type. As shown in the diagram, the effect of this action is that the funicular polygon approaches the structure in the semi arch impacted by the wind, whereas it separates from the downwind semi-arch. Therefore the downwind elements are those with the largest internal forces, because the compression force is larger, and because the funicular polygon of loads is more eccentric.

The structure on the ground floor of the building shown here is a good example of a form adapted both to carrying permanent loads and dominant variable loads. The principle is similar to that of the *Eiffel Tower*: the slope of the columns is chosen to guarantee ideal carrying of the horizontal forces generated by the wind that blows on the building, or by the inertia forces of caused by seismic movement. When the resultant of the horizontal thrusts acts approximately at the center of gravity of the building, internal forces of compression and tension are produced, which when added to the compression caused by the permanent loads generate two compression struts that remain within the sections of the columns. In spite of its form, the ground floor structure can be considered as an arch in purely structural terms.

The situation changes when the horizontal thrust acts at a lower level. Here, in fact, the tensile forces must be carried by steel reinforcement. Note that the sections of the columns vary also to adapt to this action.

As we will see later, in the upper part of the building where the columns are vertical, we also have a structure that reacts to horizontal thrusts as a frame with very eccentric funicular polygons.

UNESCO building in Paris, 1952, France, Arch. M. Breuer and B. Zehrfuss, Eng. P.L. Nervi

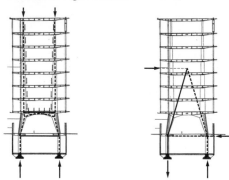

UNESCO building: transmission of internal forces induced by permanent loads and wind

Now let us consider another example in which the structure functions above all as a stabilizing element capable of carrying transverse thrusts. The steel structure of the Cachat Spring at Evian, designed by Jean Prouvé, performs this function in an optimal way. It represents a valid alternative if the structure is not to be clamped in the foundations, and when we want to avoid the diagonal elements typical of truss bracing. In this case, we have an asymmetrical three-hinged frame. One element is perfectly aligned with the resultant that passes through the hinge at the base and the one at the key, so it carries only centered forces of tension or compression when the wind blows in the opposite direction. In the other element, important eccentricities occur, and this situation is clearly shown by its ample dimensions.

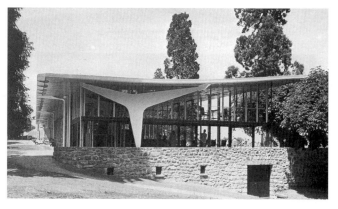

Buvette of the Cachat Spring at Evian, France, 1956, Arch. M. Novarina, constructed by J. Prouvé

Furniture in bent sheet metal, 1951, J. Prouvé

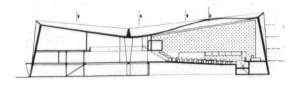

Conference room of the UNESCO complex in Paris, France, 1952,
Arch. M. Breuer and B. Zehrfuss, Eng. P.L. Nervi

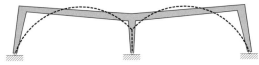

Conference room of the UNESCO complex, with the funicular
polygon of the loads

Many familiar objects can be seen as frames from a structural viewpoint. The illustrations show furniture built by Jean Prouvé by bending sheet metal. These pieces are all characterized by an adaptation of the form to structural needs.

Let us now examine a structure composed of two frames placed side by side, sharing a side wall. As shown in the diagram, the maximum eccentricity of the side walls is found in the upper part, so in those zones a larger static depth has been specified. With a corrugated surface, although the structure is relatively light, it is possible to obtain a large static depth measured between the compression and the tension zones. For the roof, the corrugated surface has been covered by a slab that undulates, following the compression zones to improve the structural efficiency and to provide, in the best position, the concrete required to carry the compression forces.

Side-by-side frames

The Parthenon, like all Greek temples, can also be considered as a frame in terms of structure. Under the influence of the permanent loads, the functioning is very simple. The architrave is composed of a series of marble blocks that behave like beams, supported by columns at their ends. These blocks are therefore subjected to both tension and compression. They are relatively squat in order to reduce the intensity of the stresses and to make up for the low tensile strength of the material. This load-carrying solution, simple but relatively inefficient, is based on the fact that the various blocks that make up the structure can transmit forces of compression to each other only by contact. A similar, monolithic structure would have a much more complex way of functioning, similar to that we have described for the conference room of the UNESCO complex in Paris.

Under the influence of the horizontal thrusts generated by wind and, more importantly, by seismic accelerations, the behavior of the structure approaches that of a frame. The resultant of the forces inside the columns can incline and counter the external thrust with its horizontal component. Stability is guaranteed as long as the resultant remains inside the section. We can observe that generally, in Greek temples, the closed structure corresponding to the *cella*, which is often quite massive, does not interact with the columns because the connection between these two structures is through the wooden parts of the roof which do not have sufficient strength and stiffness.

Although the architrave is subjected to relatively small stress, certain elements have broken over time; yet the structure, curiously enough, has in some cases not collapsed. This is due to the fact that the broken part can still function as an arch that is capable of transmitting its thrust to the ground through the inclination of the compression in the columns.

Parthenon in Athens, Greece, 477-438 BC ca., Ictinus, Callicrates and Phidias

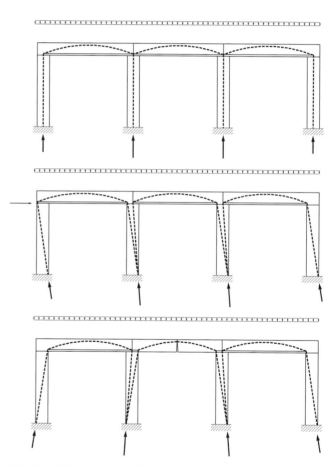

Behavior of the colonnade of a Greek temple under the influence of permanent loads, and case with a cracked architrave with the horizontal thrust of wind or an earthquake

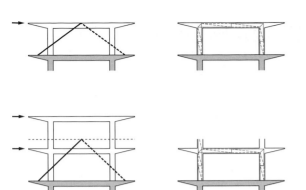

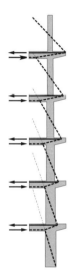

UNESCO building in Paris, internal forces of the frames on the upper levels

Now let us return to our study of the upper part of the UNESCO building in Paris, concentrating on the functioning with respect to the horizontal loads caused by wind. The structure is composed of a series of stacked frames with vertical columns. If we begin our study with the element of the upper level, we have a simple frame without any hinges. The diagram shows one possible solution, with the resultants of the compression zone downwind and those of the tension zone upwind, and the stress in the sections. The same construction can be drawn for the frames of the lower levels, all with the same structural scheme.

We can also interpret the structure as the combination of two large columns subjected to the actions of the connecting beams. In fact the beams, with their tension and compression forces exerted transversally, limit the eccentricities of the forces in the columns, thus significantly increasing the structural efficiency.

Stacked frames

UNESCO building in Paris, internal forces in the downwind columns and in one of the connection beams

Multistory side-by-side frames	The same principle was used in the Pirelli tower in Milan, Italy. Again in this case, the columns are connected to each other by horizontal beams that have the function of reducing the eccentricities of the forces in the vertical elements. In both the columns and the connection beams, the dimensions have been chosen to limit the internal tensile forces that must be carried by the steel reinforcement. This structural solution is frequently used in the construction of buildings of a certain height, both in steel and in reinforced concrete. Often a perimeter frame structure is combined with a stiffer truss structure, positioned in the central part of the building (core). The two elements of the structure each carry part of the horizontal loads, in proportion to their stiffness. A solution of this type is shown on p. 150.

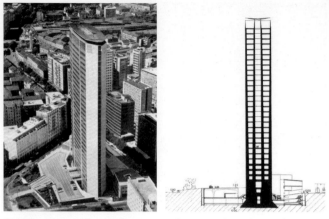

Pirelli tower in Milan, Italy, 1956, Arch. G. Ponti, A. Fornaroli, A. Rosselli, G. Valtolina, E. Dall'Orto, Eng. P.L. Nervi, and vertical section through the central columns

Vierendeel beams	Just as a multistory series of frames is capable of carrying horizontal thrusts, so a similar structure, arranged horizontally to form a beam, can carry vertical loads. This is the principle of the *Vierendeel beam*, named after the Belgian engineer who invented it toward the end of the 19th century. These beams, similar to N trusses without diagonals, actually have much more complex and much more intense stresses. In the various elements, in fact, there are forces of tension and compression with large eccentricities. Note that in order for such structures to function, their posts must be clamped into the chords, unlike trusses. As shown in the illustration, in the upper chord of a simple beam there is a compression force that is inclined with respect to the member. The forces exerted by the posts deviate the compression in the chord in a way that closely resembles what happens in the columns of a multistory frame structure.

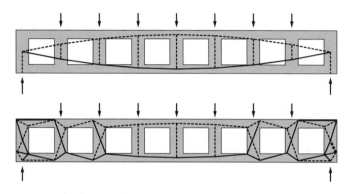

Forces in the elements of a Vierendeel beam (compression in the upper edges and tension in the lower edges) with relative eccentricities

Bridge on the Escaut river, France, 1904, Eng. A. Vierendeel

In spite of the fact that they are less efficient, in structural terms, than trusses, Vierendeel beams are often used in architecture because the rectangular openings between the chords and the posts lend themselves to the formation of doors and windows.

Villa at Jona, Switzerland, Arch. V. Bearth, A. Deplazes and D. Ladner, Eng. J. Buchli

Vierendeel beams can clearly also be stacked to increase structural efficiency. In the first example shown here, the load-carrying facade is supported by several columns in the central zone, so there are two large cantilevered consoles. In the second example, the facade with the large cantilevers covers a schoolyard. Although the loads and dimensions are large, the distribution of the forces across multiple levels and the remarkable effective depth permit limitation of the internal forces.

Project for the new municipal building in Nice, France, 2000, Arch. L. Vacchini and S. Gmühr, Eng. A. Muttoni

Léman-School in Renens, Switzerland, 2007-2009, Arch. A. Esposito, A.-C. Javet, Eng. F. Lurati, A. Muttoni, M. Felrath, M. Bosso, M. Fernández Ruiz

Deep beams
and walls

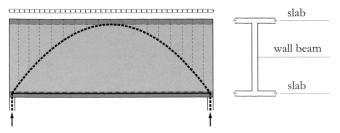

slab

wall beam

slab

Load-carrying system in a deep beam with slenderness ratio

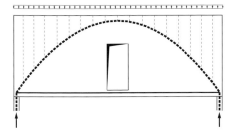

Structurally optimum positioning for openings

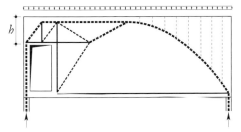

h

Load-carrying system of deep beam with openings near the supports

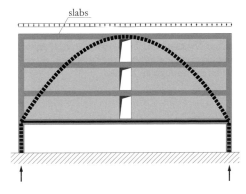

slabs

Deep wall on multiple levels to cover a large open span on the ground floor

Deep beams

When a beam fills the entire height of a floor it is called a *deep beam* or *wall*. These beams generally have very limited slenderness. If the ratio between the span and the depth is less than about three, and the supports are on the lower edge, the load-carrying system is that of an arch with tie, and true diagonals will not be formed, as in trusses, inside the wall.

Such walls often contain openings for doors and windows. From a structural viewpoint, the most advantageous position is definitely at the center, where there is a vast zone subjected to little internal force between the arch and the tie.

An opening near the supports, on the other hand, intersects the compression arch, so an alternative load-carrying system must be found. Therefore it is indispensable to provide a sufficiently strong zone above the door or under the window. The internal forces in this part will be similar to those in a reversed console, loaded by the force transmitted from the support and clamped in the deep wall.

Deep walls over several storeys

When wall beams extend for multiple floors, the structure becomes even more efficient. In these cases we talk about deep walls, using a term applied for planar structures, usually in reinforced concrete, on which the loads, the forces transmitted from the supports and the internal forces all act on the plane of the member. The efficiency of these structures is based on the large effective depth, which makes it possible to carry remarkable loads and to span very large distances (see p. 172-173). Deep walls that support buildings are often connected in a monolithic way to the horizontal parts of the construction, generally formed by reinforced concrete slabs. In this way, they can contribute to carrying the compression and tension forces in the horizontal direction.

In the example shown here, arches are formed within the deep wall, while the tie at the base can be positioned in the lower slab. This particularly efficient solution is often used when the aim is to keep the ground floor elements to a minimum in order to cover a very large, open span. Clearly the few remaining columns will have to be dimensioned to carry all the applied load, while a system of bracing also inserted on the ground floor must be capable of transmitting to the ground the horizontal thrusts generated by wind and seismic activity.

Again in this case, the definition of the minimum geometry of the deep wall must be based on study of the possible transmission of internal forces.

DEEP BEAMS AND WALLS

Folded plate structures

From a structural viewpoint, the folded plate structures we have already seen in the case of vaults and frames, or also beams, either with a wide-flange or box section, can be considered as a set of deep walls connected in a monolithic way.

The illustration shows an interesting example where concrete is used in a very efficient way.

UNESCO headquarters in Paris, France, 1952, Arch. M. Breuer and B. Zehrfuss, Eng. P.L. Nervi

Ribbed slabs, beam grids and slabs

Beam grids

By placing two series of beams perpendicular to each other, as we did with trusses, we obtain a spatial structure. The advantage of these structures, known as *beam grids*, lies in the possibility of great freedom in the positioning of the supports. We can imagine a rectangular roof with supports positioned on two sides, in which case the beams not directly resting on them function only as distribution elements, and the functioning of the structure is very similar to that of ribbed slabs.

On the other hand, we can support all the beams on the four sides, in order to carry part of the load with a series of beams in one direction and to transmit the remaining load to the supports with the other series of beams.

It is also possible to drastically reduce the number of supports, by positioning them, for example, at the four corners. In this case we are taking advantage of the structure's capacity to transmit loads in two directions, exactly like a rook, which moves in two directions on the chessboard so it can return to the corners departing from the center. As in this analogy, in which the rook can reach the corner by using many possible combinations of moves, in the beam grid the load can be transmitted along different, more or less direct routes. In beam grids with few supports we find an important increase of the internal forces in the beams that form the grid.

In the two examples shown here, a series of eight columns supports a large metal beam grid.

In the next example the grid, supported along the entire perimeter, is composed of reinforced concrete beams. They have a T section in which the flanges are wide enough to overlap and form a continuous surface.

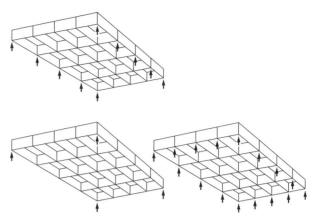

Beam grids with supports on two sides, four sides, and at the four corners

Project for the Bacardi Office Building in Santiago de Cuba, 1957, Arch. L. Mies van der Rohe, and Nationalgalerie in Berlin, 1968, Arch. L. Mies van der Rohe, Eng. H. Dienst (grid composed of wide-flange: 64.80×64.80 m, $h = 1.83$ m)

Multifunctional hall in Losone, Switzerland, 1997, Arch. L. Vacchini, Eng. R. Rossi

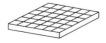

Yale University Art Gallery, New Haven, Connecticut, 1953, Arch. L.I. Kahn

In this case too, the beams are in reinforced concrete and the flange is wide enough to form a continuous surface. Here, however, three series of beams have been combined to form a triangular grid.

Slab for the Gatti woolen mill in Rome, 1951, Eng. P.L. Nervi

The next illustration shows a structure with curved ribbing, to transmit loads along the shortest route, converging radially at the supports. The ribbing around the columns, on the other hand, has the function of distributing the loads and taking up the deviation forces of the radial ribbing where it curves.

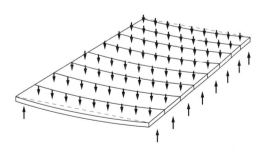

Rectangular slab supported on two sides and subject to a uniformly distributed load

Slabs

The introduction of reinforced concrete has made it possible to construct monolithic elements of large dimensions, guaranteeing the success of this structural system, which has probably become the one most commonly used in architecture today. The structural functioning of slabs is easy to understand if we imagine, within their mass, the insertion of beams and grids capable of carrying the loads and transmitting them to the supports. To better comprehend this analogy, we can examine some rectangular slabs with different load and support modes.

One-way slabs

First we will look at a slab subjected to a load uniformly distributed over its entire area and supported on two opposite sides, for example by two walls. We can cut the slab to obtain a series of strips that function like beams. If we observe carefully, we can see that they all deform in the same

RIBBED SLABS, BEAM GRIDS AND SLABS

way, and that there is no interaction between one beam and the next; the whole slab behaves in the same way.

Reinforced concrete slabs used for decks in architecture are generally subjected to permanent loads that are uniformly distributed over their surface. Even variable loads, whose intensity is usually much less than that of the permanent loads, are generally treated as distributed loads, both in the design and in the dimensioning of the structure. One exception is that of the actions exerted by walls or columns when the vertical system is interrupted along the height of the construction. In these cases, the slabs are subjected to point loads at the columns, or by loads distributed along a line, in correspondence with the walls.

Slabs are often supported by beams or beam grids that are much more rigid than the slab, due to their greater depth. The functioning and behavior of these slabs are practically identical to that of slabs supported by walls. If the beams are parallel and the controlling actions are uniformly distributed over their area, once again the slabs will work as a one-way slab.

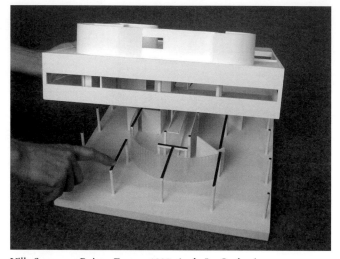

The illustration shows one of the most famous examples of this in architecture. In the *Villa Savoye* by Le Corbusier, once all the non-structural members have been removed we are left with a system of parallel beams supported by columns. The loads are first carried by one-way slabs that transmit them directly to the beams, along the shortest route.

Villa Savoye at Poissy, France, 1930, Arch. Le Corbusier and P. Jeanneret, model of the load-carrying system

Choice of slab thickness

The choice of the thickness of a slab is a typical dimensioning problem that often has to be approached already in the early phases of a project. From a structural viewpoint, as we have seen previously, any structure must be dimensioned to satisfy two criteria, defined as:
- the ultimate limit state criterion (ULS, in all cases the structure must not fail) and
- the service limit state criterion (SLS, the functioning must be guaranteed over the use of the construction).

These criteria were already known to the architects of the ancient world. According to Vitruvius, "firmitas" (solidity) and "utilitas" (functional usefulness) represent, together with "venustas" (beauty), the main qualities of a construction.

In the case of the choice of the thickness of a reinforced concrete slab, the service limit state criterion is nearly always

Reinforced concrete slab with deflection beyond the admissible limit (shelter over the ruins of Phaistos on the island of Crete)

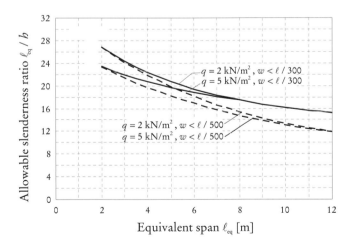

Maximum allowable slenderness ratio to satisfy the service limit state criterion in relation to the span for a rectangular slab sustained on two opposite sides by freely rotating supports. The curves indicate the different values of the admissible deflection ($\ell/300$ and $\ell/500$) and the live loads (2 and 5 kN/m²). The weight of the slab itself has been considered, as well as that of the non-structural elements, equal to 2.5 kN/m²

decisive. The thickness must be sufficient to limit deflections of the structures caused by live loads and permanent loads, because the material has viscous behavior. Deflections that are too large can, in fact, cause damage to the secondary structures such as non-structural walls, cause problems in the functioning of doors and windows, become so visible that they ruin the appearance, or prevent the drainage of water from flat roofs, as in the illustration.

Allowable deflection is usually expressed as a fraction of the span. Generally the deflection caused by permanent and variable loads should be no greater than approximatively 1/300 of the span. For a reinforced concrete slab with a span of 6 m, then, the maximum deflection tolerance is about 600/300 = 20 mm.

When deflections of the structure can cause damage to fragile elements such as doors or windows, the limit must be set more cautiously. In these cases a movement of roughly 1/500 of the span can be considered admissible, but this must be set in relation to the deformation capacity of the non-structural elements. When the limit is too restrictive, the preferable option is to add an expansion joint between the slab and the fragile elements.

Clearly, the movement also depends on the loads. The permanent loads are composed of the weight of the slab itself and that of the non-structural elements. The live loads generally vary between 2 and 5 kN/m², but they can reach much higher levels.

The diagram shows the minimum thickness of the slab in order to remain within the defined limits. As can be seen, the span plays a decisive role. In the case of a slab with a 6-m span, a modest useful load (2 kN/m²) and an ordinary requirement (deflection no greater than 1/300 of the span), we must specify a thickness of at least 0.31 m, corresponding to a slenderness ratio of 1/19.4. If we double the span we see that this limit reaches 0.78 m (1/15.3 of the span). In this case we evidently have a solution that is not very efficient, because most of the acting load is composed of the weight of the slab itself. This explains why the necessary thickness increases in a more than proportional way with respect to the span.

The influence of the live load is relatively modest. Keeping the span at 6 m and considering a variable load of 5 kN/m², the minimum thickness becomes 0.32 m (1/18.9 of the span). The influence of the admissible deflection, on the other hand, is larger. In the case of stricter requirements, with a limit equal to 1/500 of the span, we obtain a thickness of at least 0.33 m (1/18 of the span) in the case of a lightly loaded slab.

RIBBED SLABS, BEAM GRIDS AND SLABS

Influence of the type of support on slab behavior

The situation we have described corresponds to the case of a slab supported by two masonry walls, or that of two edge beams that are free to rotate. For this scheme, like the simple beams with two supports at the ends, we have an arch-cable composed of a tie and an arch that meet on the supports.

If we observe the load-carrying system of Villa Savoye we see that the slab runs on multiple beams. So we have supports that are no longer free to rotate, and the situation moves toward that of a continuous beam or a beam clamped at both ends. A similar situation can be found when the slab is attached monolithically to two reinforced concrete walls that are rigid enough to block rotation. As we have seen with beams, in these cases the stresses are reduced. The deflection is also affected beneficially, thanks to the prevention of rotation, so it is even reduced to 1/5 of what could be seen in a slab left free to rotate. Because the deflection depends on the span to the fourth power, a deflection of 1/5 can be seen in a slab free to rotate with a span reduced by a factor of 0.67, as seen in the diagram.

A continuous or clamped slab with a span 6 m will therefore be deformed exactly like a slab with a span of 4 m free to rotate. This means we can choose the thickness of the slab required to satisfy the service limit state criterion by using the diagram on the previous page and considering an equivalent reduced span. With modest useful load and an ordinary requirement we will have to provide for a thickness of at least 0.18 m (1/22.4 of the equivalent span). As we have seen, clamping permits a significant reduction of the minimum thickness of the slab.

If the slab is clamped on just one side while the opposite support remains free to rotate, we will have a situation between the two cases examined above, so we can consider an equivalent span ℓ_{eq} equal to 0.8 times the effective span.

In the case of a cantilevers, not only the internal force, but also the deflections become very important. In this case the equivalent span corresponds to 1.76 times the overhang of the cantilevers. A cantilevers with an overhang of 6 m will therefore require a thickness corresponding to that of a slab simply placed over a span of 10.56 m. We obtain a thickness of as much as 0.66 m (1/16 of the equivalent span), so the structure is far from efficient. In these cases we need to choose a more suitable section, or increase the stiffness of the slab by means of prestressing.

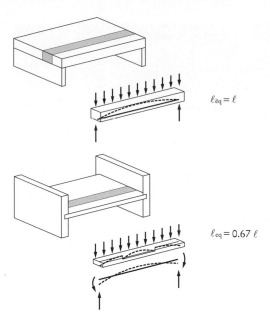

$\ell_{eq} = \ell$

$\ell_{eq} = 0.67\,\ell$

Comparison between a slab clamped on two sides and one on simple supports; arch-cable

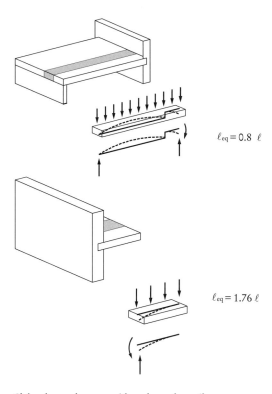

$\ell_{eq} = 0.8\ \ell$

$\ell_{eq} = 1.76\,\ell$

Slabs clamped on one side only, and cantilevers

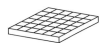

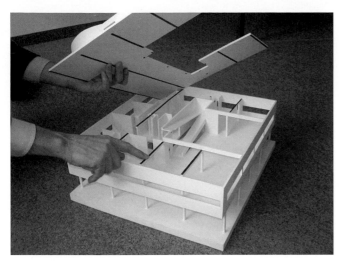

Villa Savoye at Poissy, France, 1930, Arch. Le Corbusier and P. Jeanneret, model of the load-carrying structure with the slab

The considerations made thus far apply only to loads uniformly distributed over the surface, which can be transmitted in only one direction toward the supports located on two opposite sides of a rectangular slab. It should also be noted that in these cases, when we talk about span we mean the distance between the supports.

Continuous slabs

The illustration shows a situation that is very different in geometric and static terms. Where there is continuity between slabs with comparable spans, we can hypothesize a clamp, while when the slab is supported at the edge by walls or beams, we have behavior similar to that of a support that is free to rotate.

To understand the functioning, it is useful to isolate the different elements, identify the type of support and interpret the directions along which the loads can be transmitted to the supports.

Functioning with concentrated loads

Now we consider loading the slab supported on two sides with a concentrated force at the center. If we cut the slab into strips, as in the earlier case, we have one loaded beam that will tend to lower, while the others remain in their original state without deformation. This certainly does not correspond to the real behavior of the slab. In fact, a grid is established, where the crossbeams near the concentrated load play the role of distributing the load to multiple strips in the other direction. This means that a concentrated load is actually carried in all directions.

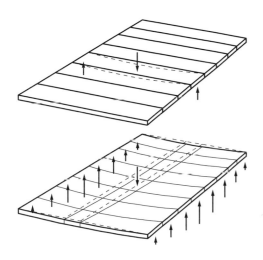

Rectangular slab supported on two sides and subjected to a concentrated load

Two-way slabs

In a rectangular slab supported on four sides, strips can be established in multiple directions also under the effect of a uniformly distributed load. The difference with respect to the grid lies simply in the fact that in the slabs, the directions of the beams we hypothesize to describe the functioning are not pre-set, but follow the most efficient path to transmit the loads to the supports.

As we will see later, this greater efficiency is reflected in reduced internal forces and, above all, smaller deflections.

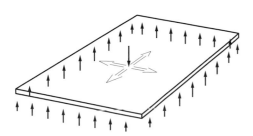

Rectangular slab supported on four sides

RIBBED SLABS, BEAM GRIDS AND SLABS

Equivalent spans for two-way slabs

The chart shows the equivalent spans for different support modes. The first chart refers to a square slab, while the second shows the values for a rectangular slab, with a ratio of 2/3 between the small and large spans.

When a square slab is supported in the same way on all four sides, half the load is transmitted in one direction and the other half is transmitted in the other direction. The slab will be subjected to smaller internal forces than one that functions in only one direction, and will undergo smaller deflections. This advantage is reflected in the decrease of the equivalent span, meaning that the minimum admissible thickness can also be reduced. A square slab with free supports on all four sides, for example, has an equivalent span equal to 0.75 times the effective span. If we consider a square slab with 6-m sides, then, we obtain an equivalent span that is reduced to 4.5 m. So we can choose a minimum thickness of 0.21 m (1/21.6 of he equivalent span), with respect to the 0.31 m calculated for the span with two supports.

If, on the other hand, we examine a slab measuring 6 × 9 m, we obtain an equivalent span of 0.88 times the smaller span, so the necessary thickness rises to 0.26 m (1/20.5 of the equivalent span). Therefore beneficial effect of the transmission of loads in two directions is significantly less than in the case of the square slab. This advantage becomes practically negligible when the longer span is more than twice as large as the smaller span. This means that with these geometries the minimum thickness can be estimated by considering the equivalent span of the one-way slabs.

Equivalent span, expressed as the ratio ℓ_{eq}/ℓ, for square slabs with different support modes and uniformly distributed load (… free edge, – supported edge, = clamped edge)

Equivalent span, expressed as the ratio ℓ_{eq}/ℓ_{min}, for rectangular slabs with ratio 2/3 between short and long spans, and different support modes (… free edge, – supported edge, = clamped edge)

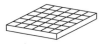

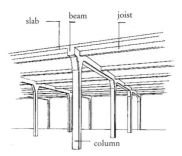

Explanatory drawing of the patent "Système Hennebique" for floor/roof slabs in reinforced concrete, 1898, Eng. F. Hennebique

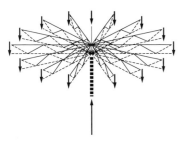

Functioning of the slab-column joint

Slabs on columns

In the first reinforced concrete slabs supported by columns, beams were placed on the intrados to distribute the effect of the force transmitted by the supports. There was thus a combination of a slab and a beam grid that imitated the earlier constructions in wood and steel.

One of the first to understand the enormous potential of reinforced concrete slabs was R. Maillart at the beginning of the 20th century, who patented a system of slabs supported by columns without the use of beams for reinforcement. Its functioning is identical to that of the slab with curved ribbing constructed by P.L. Nervi, in which the ribs are arranged radially, extending from the column (see p. 205). In fact beams of this type do form inside the slab. Their internal forces, similar to those of a continuous beam in the zone of the supports, are represented in the diagram.

Mushroom slabs

The convergence of all the beams in a relatively small zone causes a very intense concentration of internal forces in the lower zone under compression, the middle zone, where diagonals in compression and tension form, and in the upper zone, in tension.

To resolve this problem Maillart increased the thickness of the slab in the critical zone, obtaining a slab known as the "mushroom slab".

The famous "mushrooms" built by F.L. Wright have an identical way of functioning. This example, more than others perhaps, illustrates the efficiency of a structure of this type.

Mushroom slab at Giesshübelstrasse, Zurich, Switzerland, 1910, Eng. R. Maillart

Johnson and Son administration building, Racine, Wisconsin, USA, 1939, Arch. F.L. Wright

Flat slabs

The use of stronger concrete, suitable reinforcement and, above all, the choice of a sufficiently thick slab make it possible to avoid mushroom capitals on columns. A slab thus supported has very large stresses near the columns, while the internal forces in the span are relatively small.

Flat floor slab supported by columns in the Visual Arts Center, Cambridge, UK, Mass., 1964, Arch. Le Corbusier

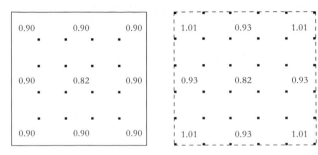

0.90	0.90	0.90
0.90	0.82	0.90
0.90	0.90	0.90

1.01	0.93	1.01
0.93	0.82	0.93
1.01	0.93	1.01

Equivalent spans for some cases of flat floor/roof slabs supported by columns

In reinforced concrete slabs supported by columns, the beneficial effect of the transmission of loads in two directions is countered by the concentration of forces near the supports. In the choice of thickness, in fact, the ultimate limit state criterion can often become decisive.

When the critical zone is strengthened by mushrooms capitals or particular reinforcements and the serviceability limit state controls the design, the thickness of the slab can be estimated by using the equivalent spans shown here, and using the diagram on p. 207.

It can be observed that in flat slabs with a regular grid, the edge spans have the largest deflections and become decisive when the thickness is kept constant throughout the slab. When the slab is supported on columns along the edge also, the equivalent span practically corresponds to the distance between the columns.

Choice of thickness of slabs supported by columns

Stability of compressed members

We have already repeatedly seen that in the design of a structure we always have to discern whether the internal forces are of tension or of compression. The identification of these different forces is important for two reasons.

First of all, certain materials like concrete, stone and masonry, for example, are highly resistant to compression, while their tensile strength is limited. Furthermore, the transmission of a compression force from one element to another is usually much simpler, because mere contact is sufficient.

On the other hand, as we have already emphasized, the carrying of compression forces has its own problems. If a tensile force in a structure causes deformations and deflections that make the structure approach the line of action of the internal force, the same structure, if subjected to compression, will tend to move away from the ideal position, so the internal forces will intensify. This is true for a series of rods connected by hinges, and for a monolithic structure.

In the first case, we have an unstable structure. As shown in the illustration, if the system is subjected to tension, the rods that form a mechanism move until they coincide with the funicular polygon of loads, notwithstanding the initial eccentricity. As soon as the eccentricity is eliminated, even an unstable structure will be capable of bearing expected loads, as long as the material is strong enough.

On the contrary, when the system is subjected to compression, even the smallest initial eccentricity can suffice to make the rods move away from the position of equilibrium. A structure of this type is, effectively, unstable.

All this is easy to grasp intuitively by looking at the example in the illustration, in which the funicular polygon is a straight line, but it applies in all cases. On p. 73 we have already examined a system in which the funicular polygon has a trapezoidal form, and we reached the same conclusion: a cable is stable, while an arch is not.

Unstable mechanism of rods under tension and the same system under compression

In order to be sufficiently stable and to resist compression, a column must then be capable of carrying a transverse load without too much deformation.

We can analyze our column by knocking it down, treating it as if it were a beam loaded at the middle, and studying its stiffness.

Therefore to study the stiffness and stability of a column, we have to lie it down and load it at mid-span.

We should not forget that the internal forces in the beam are similar, but not identical, to those of our column sujected to compression only. This analogy is useful only to study the parameters that influence stiffness, so as to be able to propose counter measures to avoid instability.

As we have already seen in our coverage of beams, these parameters are:

1. the span of the beam (in our case, the height of the column);
2. the constraints (in other words, how the ends are held in place);
3. the stiffness of the material;
4. the dimensions of the section;
5. the shape of the section.

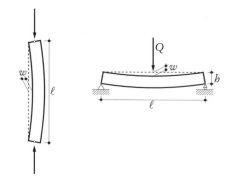

Analogy between an unstable column and a beam with transverse load

Influence of column height on compressive strength

The influence of length on the bending stiffness is quite clear. Keeping all the other parameters constant, a beam with a large span is much less rigid than a short beam. This means that tall columns are much more subject to the phenomenon of buckling than short columns. In the graph, the critical load of the rod we have described is expressed in relation to its length. As we can see, the strength, equal to just 40 N when the length is 800 mm, can decrease further when the length increases. We can also see how the strength of the material can be completely exploited only when the length is less than about 20 mm; to be able to utilize at least half the strength of the material, the length should not be more than 100 mm.

We can generalize from these data, stating that in a steel column with a full circular section and a yield strength $f_y = 235$ N/mm², the strength of the material is completely utilized when the height is less than five times the diameter, and this ratio must not go beyond 25:1, if we want to use at least 50% of the strength of the material.

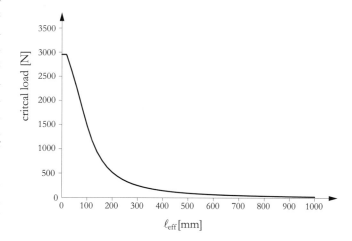

Influence of the length of the column on compressive strength (critical load)

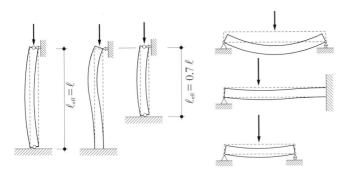

Comparison of buckling modes of a column free to rotate, one clamped at the base, and one 70% of the length; analogy with a beam with comparison of deflections (stiffness)

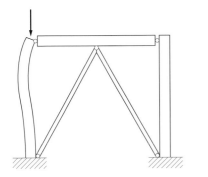

Example of a column clamped at the base and free to rotate at the top

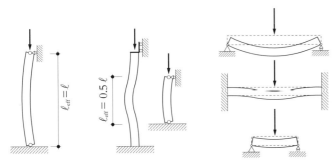

Comparison of buckling modes of a column free to rotate, one clamped at the ends, and one 50% of the length; analogy with a beam with comparison of deflections (stiffness)

The relationships between slenderness, seen as the ratio between the height of the column and its diameter, and the degree of utilization of the material, which we have just described, also apply to the model of the column being analyzed, assuming that the ends are attached but free to rotate.

If the end is clamped, instead, preventing rotation, the critical load increases. For our model with a column height of 800 mm, the critical load is almost doubled, from 40 to 77 N. The increase is easy to explain, considering analogy with the beam. If one end is clamped, the same transverse load will cause smaller movements. In other words, the bending stiffness will be increased.

The new column has a critical load comparable to that of a shorter column, but with the ends free to rotate. For our model, we have to reduce the height from the previous 800 mm to 560 mm to keep the critical load equal to 77 N. This new length corresponds to 70% of the initial length, and is known as the effective length ℓ_{cr}.

The image shows a typical structural scheme, where the column has the constraints described. The clamp at the base prevents the column from rotating with respect to the foundation, while at the top a hinge between the column and the horizontal beam makes it free. In that point, however, the column is not free to move horizontally due to the presence of a bracing system.

When both ends of the column are clamped the stiffness and, therefore, also the critical load are even greater. Naturally the upper end, although clamped, must be free to move vertically. Otherwise the load would not be carried by the column because it would be completely carried by the support.

Our 800 mm column now has a critical load of 147 N. To obtain the same critical load effective with the column free to rotate at the ends, we would have to reduce its height to 400 mm. In other words, for the column clamped at the two ends the effective length corresponds to 50% of the total height.

We encounter these columns in frames, where the horizontal element is much more stiff than the column, so the rotations at the ends are prevented. To make horizontal displacements at the top impossible, again in this case we need a rigid bracing system. The columns of the *Carré d'Art in Nimes*, which look very slender, are clamped to much stiffer upper beams. The horizontal movement of the beams, and therefore also of the top of the columns, is prevented by a horizontal bracing system attached to the building.

If the bracing is missing, the frame becomes less rigid, because it can move horizontally. The buckling mode also changes radically: the largest movement happens at the upper end.

Our 800 mm column will have, once again, a critical load of 40 N, identical to that of the column left free to rotate at its ends. This means that in these columns too, the effective length corresponds to the height of the column.

Based on this consideration, we can state that bracing, namely a structural system designed to carry horizontal loads, also has another important function: that of increasing the critical load of the columns and therefore their strength with respect to vertical loads.

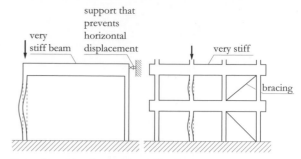

Examples with columns clamped at the ends

Carré d'Art in Nimes, France, 1993, Arch. N. Foster, Eng. Studio Ove Arup + OTH Mediterranée

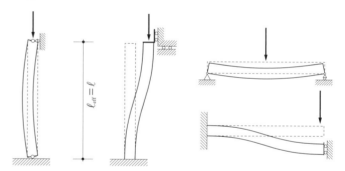

Comparison of buckling modes of a column left free to rotate and one clamped at the ends but free to move in a horizontal direction; analogy with a beam, with comparison of deflections (stiffness)

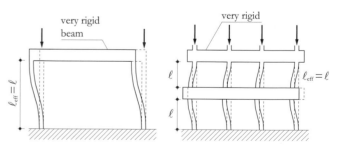

Examples with columns clamped at the ends but free to move horizontally

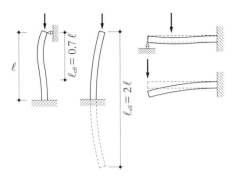

Comparison of buckling modes of a column left free to rotate but constrained at the ends, and a column clamped at the base and completely free at the top; analogy with a beam, with comparison of deflections (stiffness)

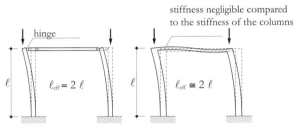

Examples with columns clamped at the base and completely free at the top

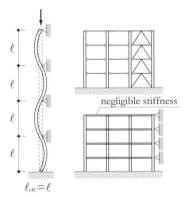

Buckling mode of a continuous column and frame with critical length of the column equivalent to floor height

Columns of Georges Pompidou Centre in Paris (see p. 9 and p. 139)

If we eliminate the bracing system in the case of a column left free to rotate at the top, thus making it free to move horizontally, we obtain a significant reduction of the critical load. The strength of our model with the rod 800 mm high and 4 mm in diameter is reduced from the initial 77 N to just 10 N.

To obtain the same reduction with a column left free to rotate but restrained horizontally we would have to increase its length from the initial 800 mm to 1600 mm. This means that the effective length corresponds to twice the height of the column.

Finally, let us consider the case of a multistory frame with continuous columns, that are left free to rotate, because they are connected to the horizontal beams by joints. Given the fact that these joints can transmit horizontal internal forces, the columns are not capable of movement in that direction.

A very similar situation arises when, in a frame, the horizontal beams are clamped to the columns but have much lower bending stiffness than the columns. In this case, they are not able to prevent rotation of the columns.

In these examples the columns can buckle, forming a wave that has an effective length that corresponds to the height of the floors. The critical load, then, is identical to that of the columns if they were interrupted at each floor.

One example is that of the columns of *Georges Pompidou Centre* in Paris. At every floor, the columns are constrained by a bracing system, but they are free to rotate.

As we have seen, the constraints have an important influence on the bending stiffness and, therefore, on the critical load of the columns. This phenomenon can be aptly described in terms of the effective length, which corresponds to the height of a column hinged at the ends, which has the same critical load.

Every time we constrain a column, with the aim of preventing horizontal displacement or rotation, we decrease its effective length and therefore also increase its critical load.

STABILITY OF COMPRESSED MEMBERS

The graph shows the critical load of our column model in relation to height, with the various possible constraints and corresponding effective lengths.

Influence of the stiffness of the material on the critical load of a column

A material with a low modulus elasticity of will clearly be more likely to buckle. We have already seen such a situation, with the block of foam rubber.

This particular factor is important for aluminium, whose elasticity modulus is one-third that of steel. So we can state that aluminium members under compression are particularly delicate in terms of stability.

The next graph shows the critical load of our model with the steel rod, compared with an identical model with an aluminium rod.

When we compare different materials, it is above all the modulus of elasticity that influences the critical load of slender columns. The strength of the material, on the other hand, has a negligible influence, and becomes determinant only for thick, block-like columns.

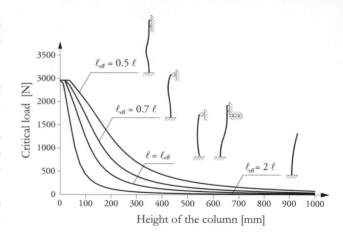

Critical load in relation to the height of the column with the different possible constraints (S235 steel, diameter 4 mm)

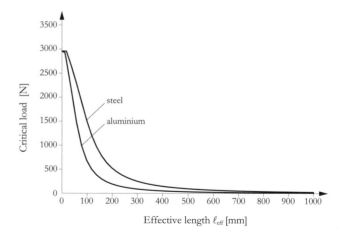

Critical load in relation to the effective lenght for a rod in S235 steel and one in aluminium with identical yield stress

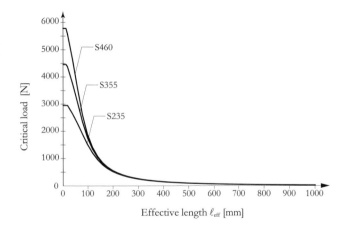

Critical load in relation to the effective length for three rods in S235, S355 and S460 steel (diameter 4 mm, columns free to rotate at the ends)

This consideration has a very important practical application: it makes no sense to use high-strength steel when the columns are too slender to make use of the strength of the material. This is shown very clearly in the diagram, in which the critical load of our model, with three rods in S235, S355 and S460 steel, is plotted in relation to the slenderness of the column.

Other construction materials, like concrete, masonry or wood, for example, have a much lower modulus elasticity of than steel. To compensate for this, their sections are usually much larger, so stability, although certainly not a factor to be neglected, is nevertheless a less crucial problem than in the case of steel structures.

In a steel column, too, one fundamental parameter that influences stability and strength is the dimensions of its section. An increase does not augment only the area but also its bending stiffness, and therefore its stability. So we have the overlapping of two effects: a larger available area and reduced slenderness, so the material can be more fully exploited.

To quantify this effect, let us return to our example with a rod 4 mm in diameter with effective length equal to 800 mm. As we have seen, the critical load corresponds to 40 N, with utilization of the strength of material equal to 1%. If we double the diameter, so that the area of the available material is quadrupled, and keep the effective length constant, the strength increases to 585 N, or 15 times the initial strength. The level of utilization rises to 5%.

We can double the diameter again and observe that the strength increases even more rapidly with respect to the increase in area (see the chart).

With $\phi = 32$ mm we have an area that is 64 times that of the original. However the critical load, is 2500 times that of the rod with $\phi = 4$ mm, with a degree of utilization that reaches 50%. This degree of strength utilization is, in fact, the same as the one we saw with a column that has a slenderness ratio of 25 (100 mm / 4 mm or, as in this case, 800 mm / 32 mm).

Influence of the dimensions of the section

Diameter ϕ [mm]	Area [mm²]	Increase of area	Critical load [N]	Increase of critical load	Utilization of the material
4	12.57		40		1%
8	50.27	× 4	585	× 15	5%
16	201.06	× 16	8328	× 215	18%
32	804.25	× 64	95222	× 2500	50%

Influence of the dimensioning of the column on the critical load (effective length = 800 mm, S235 steel)

Influence of the shape of the section

To increase the bending stiffness of a column and, therefore, also its capacity to resist buckling, a much more efficient system exists with respect to that of simply increasing the dimensions by increasing the diameter.

As we have seen for the stiffness of beams (see p. 172), a solid rectangular section is much less efficient than a wide-flange section with an equal quantity of material and the same depth. In general, we can say that the larger the distance between the compression and tension zones, the larger the bending stiffness of the column and, therefore, its capacity to prevent buckling.

The efficiency of a column also depends, then, on the shape of its section, which plays a very important role. So it is of fundamental importance to make intelligent use of the material, using sections with slender walls that have a large distance between the tension and compression zones.

The graph shows a comparison of wide-flange, box, tubular and solid sections, with the same area, and therefore with the same quantity of material used and a maximum bulk of 0.30 m.

It is clear that a solid section is much less efficient than a tubular, box or wide-flange section.

As a result, the choice of the type of section must also consider the type of internal forces. When there is tension, as we have seen, the shape does not influence the strength of the section, so we can also choose solid or compact sections. For members in compression, if we want to correctly utilize the material, we should choose hollow or wide-flange sections.

It is important to point out here that wide-flange sections, unlike box and tubular sections, have different stiffnesses in the two directions. As we have already seen, a wide-flange beam with its flanges positioned vertically is much less stiff than the same beam positioned in a conventional way with the flanges horizontal.

Therefore in members under compression with a wide-flange section it is useful to limit the critical length by stabilizing elements that constrain the column and prevent it from moving in the direction of its lesser stiffness. In the other directions these elements are unnecessary, or they can be positioned at a greater distance, because a larger effective length is not a governing factor.

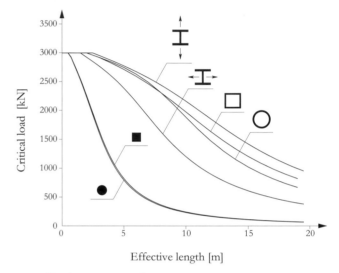

Critical load: comparison of sections

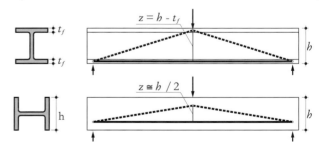

Stiffness of a beam with wide-flange section: flanges arranged horizontally and vertically

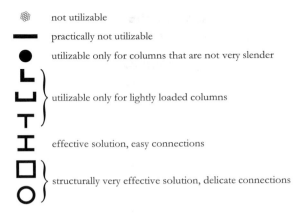

● not utilizable

▬ practically not utilizable

I utilizable only for columns that are not very slender

⊔ ⏉ ⬡ } utilizable only for lightly loaded columns

⬚ effective solution, easy connections

○ } structurally very effective solution, delicate connections

Choice of steel sections for members under compression

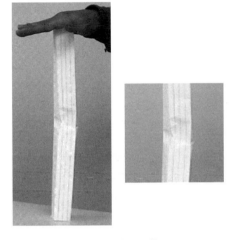

Local buckling of a column with slender walls

Risk of   Ideal section   Risk of local
buckling   buckling

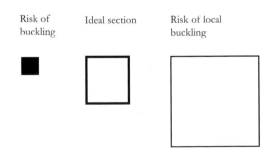

Sections with the same amount of material

As a result, the choice of the type of section must be based on analysis of the internal forces. The chart shows the sections most often used in compressions steel members. These considerations do not apply to elements in tension, for which only the area of the section and the yield strength of the steel influence the strength.

For structures in reinforced concrete, wood or masonry, from a certain degree of slenderness it also becomes important to choose efficient sections.

The rule stating that in order to prevent buckling of a column it is best to use sections with slender walls, with a large distance between the tension and compression zones, nevertheless has its limits. In fact, when the box or tubular column whose cross-section has very slender walls is loaded, before it buckles completely a phenomenon of local instability happens in the slender walls, known as *local buckling*.

This can easily be simulated by building a model of a tubular-section column by rolling up a sheet of paper. When a load is applied to the model, we can see a sort of "local buckling" of "small columns" formed inside the slender walls, before the column buckles completely.

In the design of the section of a column it is necessary to find the proper compromise between a solid, full section, with the risk of buckling, and a section with walls that are too slender, which can cause local instability. The optimum section of a box-section column has a ratio between the width of the column and the thickness of its wall ranging from 30 to 50.

Choice of sections

Local buckling

Trusses and Vierendeel columns

When it is important to reduce the quantity of material, it is preferable to use trusses or Vierendeel beams for members in compression. To guarantee the necessary stability in all directions, these elements must have at least three chords. These solutions are often used because they also guarantee great transparency.

Vierendeel beams as compression elements in the Millennium Dome, UK, 1999, Arch. R. Rogers, Eng. Studio Happold

Variable-section columns

Variation of the section is often used to achieve the same goal and obtain a structure that looks more slender. As in the two examples, in the zones where the internal forces are largest a stiffer section is chosen, while in the other zones where the eccentricity with respect to the funicular polygon of the loads is smaller and stiffness is not as necessary, the dimensions of the member can be drastically diminished.

These solutions also make an elegant transition possible between the central zones, with larger sections, and the jointed ends in which the force has to pass through the hinges.

For a definition of the optimal form of compression elements we can make use of the analogy with the problem of carrying transverse loads. Thus we can apply a virtual load and construct a corresponding arch-cable, which will directly supply us with the desired form.

The Bigo in Genoa for the Columbian Exhibition, Italy, 1988-1992, Arch. R. Piano, Eng. P. Rice (Studio Ove Arup)

Palazzetto dello Sport in Rome, Italy, 1957, Arch. + Eng. P.L. Nervi

Appendices

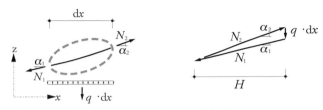

Infinitesimal cable segment and polygon of the forces

First we will examine the case of a cable with a distributed load of constant intensity (see p. 47). As we saw on p. 72, the case of an arch can easily be derived by analogy, if we consider a free body diagram that includes a segment of cable of infinitesimal length, as shown here. Internal forces N_1 and N_2 are applied on the two ends, with different slopes, because the load applied to the infinitesimal part has caused a deviation of the cable. As in all the examples with vertical loads, the horizontal component of the force vectors is constant, and corresponds to the horizontal force on the supports H.

The condition of equilibrium in the infinitesimal free-body leads to

$$H \cdot (\tan \alpha_2 - \tan \alpha_1) = q \cdot dx$$

Expressing $\tan \alpha_1$ with the first derivative of the function that generates the curve : $\tan \alpha_1 = z'$
and $\tan \alpha_2$ with the increase of slope $\tan \alpha_2 = z' + z'' \cdot dx$

we obtain:

$$H \cdot (z' + z'' \cdot dx - z') = q \cdot dx$$

from which

$$z'' = q/H.$$

The function $z(x)$ that describes the funicular curve can be determined by integrating the constant q/H two times:

$$z' = q/H \cdot x + C_1$$
$$\text{and } z = q/H \cdot x^2/2 + C_1 \cdot x + C_2$$

where C_1 and C_2 are constants of integration and are determined by considering the geometric conditions at the supports. The function obtained is that of a second-order parabola. When the cables are loaded only by their own weight and the section of the cable is constant, we have to consider that the weight of the element is

$$q \cdot ds$$

where $ds = \sqrt{dx^2 + dz^2} = dx \cdot \sqrt{1 + z'^2}$ is the length of the infinitesimal part (in the previous case the load was equal to $q \cdot dx$). So the differential equation becomes

**Appendix 1.
Analytical
determination
of the funicular
curve with
distributed load**

APPENDICES

$$z'' \quad \frac{q}{H} \cdot \sqrt{1+z'^2}$$

and the solution is that of the catenary curve

$$z = C_1 \cdot \cosh C_2 x = C_1 \cdot \frac{e^{-c_2 x} + e^{c_2 x}}{2}$$

where C_1 et C_2 are two constants.

Appendix 2. Analytical expression of the equilibrium conditions

The first equilibrium condition, according to which "the forces that act on a free body are in equilibrium if they vectorially cancel each other out" (see p. 11 and 30) can be formulated as follows:

$$\sum \vec{F} = 0$$

where the symbol $\vec{F}$ indicates that the forces F are actually vectors. If for every acting force we consider the components according to a defined system of axes, the previous equation can be replaced by the following system:

$$\sum F_x = 0$$
$$\sum F_y = 0$$
$$\sum F_z = 0$$

As we have seen, the condition of equilibrium just described is necessary but not sufficient. We have also found a second condition of equilibrium: in the case of two forces, they must be on the same line of action, while in the case of three forces the three respective lines of action must meet in one single point (see p. 30).

To analytically describe these conditions, which are actually equivalent, it is useful to introduce the notion of moment. The moment of a force with respect to a point of reference is defined as the product of the force times the distance between the point of reference and the line of action of the force. The moment, too, is really a vector that results from the vectorial product of the force and a vector $\vec{r}_A$ which indicates the segment between the point of reference A and any point along the line of action of the force:

$$\vec{M}_A = \vec{r}_A \times \vec{F}$$

When all the forces act on a plane that also contains the point of reference A, all the moments are vectors perpendicular to

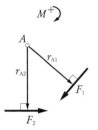

The moment (of a force with respect to a point)

the plane itself. The intensity of the moment corresponds, then, to the product of the intensity of the force F and the distance r_A:

$$M_A = r_A \cdot F$$

In this case, however, it is necessary to define the direction in which the forces create a positive moment (implicitly the direction of the axis perpendicular to the plane on which the forces are located). As seen in the example, the moment of the force F_1 with respect to point A is positive, while the moment of force F_2 is negative

The second condition of equilibrium can thus be formulated in the following way: "a free body is in equilibrium only if the sum of the moments caused by the forces with respect to any point of reference A is canceled out". This means that with $M_{F,A}$ the moment caused by each force F with respect to point A:

$$\sum \vec{M}_{F,A} = 0$$

or, if all the forces are on one plane:

$$\sum M_{F,A} = 0$$

Note that this condition of equilibrium, thus formulated, applies not only in the case of two or three forces, but also for any number of forces.

These analytical expressions of the conditions of equilibrium are very useful to find the resultant of parallel forces or support forces of systems with parallel loads.

In the case of the structure shown in the illustration (see p. 130), the second condition of equilibrium makes it possible to rapidly find the force acting on the support to the right. If we take as a point of reference a point located at the left support, the two components of the force transmitted by that support, as yet unknown, do not cause any moment and therefore are not part of the equation that makes it immediately possible to find the force F_B on the right support:

$$\sum M_{F,A} = 0 = 1\,[\text{m}] \cdot 30\,[\text{N}] + 3\,[\text{m}] \cdot 10\,[\text{N}] - 4\,[\text{m}] \cdot F_B$$

so that $F_B = 15\,[\text{N}]$.

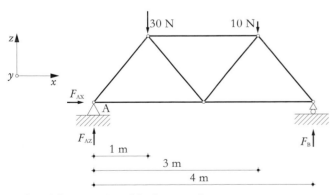

Analytical determination of the forces on the supports

Now we can easily find the forces on the left support, using the first condition of equilibrium. The horizontal component is derived directly from:

$$\Sigma F_x = 0 = F_{Ax}$$

while the vertical component can be obtained from

$$\Sigma F_z = 0 = F_{Az} - 30\,[N] - 10\,[N] + F_B$$
$$= F_{Az} - 30\,[N] - 10\,[N] + 15\,[N]$$

for which $F_{Az} = 25\,[N]$.

The resultant of the loads can be found in the same way. In the diagram with two vertical loads, the resultant R must also be vertical, obtained from

$$\Sigma F_z = 0 = R - 30\,[N] - 10\,[N]$$

and thus $R = 40$ N. The position of the line of action of the resultant (distance r_R from the point of reference A) can be determined by

$$\Sigma M_{F,A} = 0 = 1\,[m] \cdot 30\,[N] - r_R \cdot R + 3\,[m] \cdot 10\,[N]$$

$$= 1\,[m] \cdot 30\,[N] - r_R \cdot 40\,[N] + 3\,[m] \cdot 10\,[N]$$

for which $r_R = 1{,}5\,[m]$.

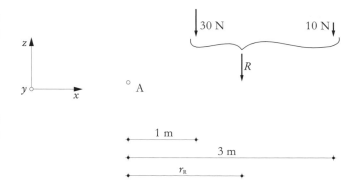

Analytical determination of the resultant of the loads

Appendix 3. Axial force, shear force and bending moment

In statics, the resultant of the acting internal forces where a free body has been isolated is often indicated by means of its components according to a longitudinal axis and two transverse axes. When the structures and loads are in one plane, there will be only one transverse axis, it too on the principal plane, and perpendicular to the longitudinal axis. Thus the longitudinal component of the resultant of the internal forces is defined as *axial force N*, and the transverse component as *shear force V* (see p. 137).

Since the resultant of the internal forces does not necessarily passes through the axis of reference (in the case of beams the barycentric axis is often used as the reference), we need to define another magnitude called bending moment *M*, corresponding to the moment of the resultant of the internal forces with respect to the meeting point between the axis of reference and the limit of the free body.

In the example shown on p. 133 the axial force, the shear force and the bending moment represent the components of

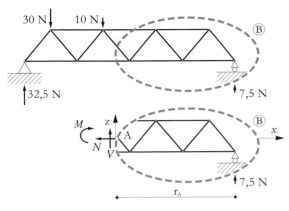

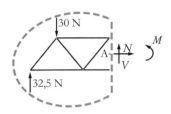

Determination of the internal forces N, V and M (see p. 135)

Determination of the forces N, V and M (see p. 136)

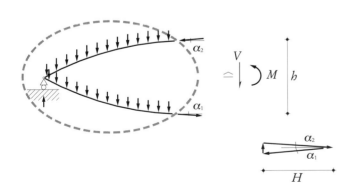

Determination of the forces in an arch-cable

the total resultant of the force acting on the truss (upper chord, lower chord and diagonal cut by the free-body diagram shown here). The conditions of equilibrium immediately give us the forces we seek:

$$\sum F_x = 0 = N$$ (x in this case is the axis of reference of the structure)

$$\sum F_z = 0 = V + 7{,}5 [N]$$ from which we obtain
$$V = -7{,}5 \ [N]$$

$$\sum M_{F,A} = 0 = M - r_A \cdot 7{,}5 [N]$$ from which we obtain
$$M = r_A \cdot 7{,}5 \ [Nm]$$

Note that in the right part of the truss the bending moment corresponds to the product of the support force (7.5 N) times the distance from the support r_A. The linear shape of the bending moment in this resulting zone is clearly visible in the illustration on p. 134.

The procedure can clearly also be utilized in the case of multiple forces acting on the free body. In the example shown here (see p. 136) we have:

$$\sum F_x = 0 = N$$

$$\sum F_z = 0 = 32{,}5 [N] - 30 [N] - V \quad \text{from which we obtain}$$
$$V = 2{,}5 \ [N]$$

$$\sum M_{F,A} = 0 = 3 \cdot 32{,}5 - 2 \cdot 30 - M \quad \text{from which we obtain}$$
$$M = 37{,}5 \ [Nm]$$

Starting with the internal forces N, V and M, we can determine the forces in the arch and the cable of arch-cable, in the chords and diagonals of trusses, as well as the internal forces of beams and frames.

As we have already seen at p. 134, the shape of the moments resembles that of a arch-cable. In fact, as shown in the illustration here, we have

$$M = H \cdot h$$

where H, in the case of vertical loads, is constant throughout the length of the structure. The shear force V, furthermore, coincides with the vertical component of the forces in the arch and the cable:

$$V = H \cdot \left(\tan \alpha_1 + \tan \alpha_2 \right)$$

On p. 137 we have already seen how the force in the diagonal of a truss with parallel chords can be directly determined by starting with the shear force V. The bending moment M, on the other hand, influences the force in the chords (see the illustrations here and on p. 134):

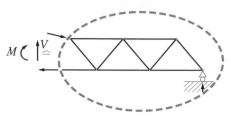

Determination of the forces in a truss

$$N_{chord} = \pm \frac{M}{h}$$

where the + sign applies to the lower chord and the − sign applies to the upper chord. To be precise, this relationship only applies when the limit of the free body intersects the node opposite the chord under consideration. Otherwise, we have to also account for the shear force on the chords.

For beams, we have to consider that the effective depth z between the compression and the tension zone is less than the height of the section (see p. 172). With this correction, the relationship just described remains valid:

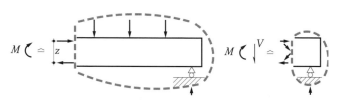

Determination of the forces in a beam

$$N_{chord} = \pm \frac{M}{z}$$

Glossary

Abscissa	The horizontal axis of a graph.	**Area**	The surface of the cross section. See *section*.
Action	Imposed force or deformation affecting a structure.	**Arris**	The line of intersection between two surfaces not on the same plane.
Adimensional	Indicates a pure numeric magnitude, without units of measure.	**Atmosphere**	A unit of pressure, corresponding to about 10^5 [N/m^2].
Affine	An object is said to be affine with respect to another if they are related by a geometric transformation (the affine transformation) that preserves collinearity and ratios of distances.	**Auxiliary arch or cable**	An auxiliary graphic construction that permits determination of the position of the resultant of multiple forces.
Aggregate	Combination of inert materials used for the composition of mortar and concrete.	**Axial force**	An internal force of tension or compression acting perpendicular to the section of a structural element.
Allowable deflection	The maximum allowed displacement or deformation of a structural element, i.e., one that does not compromise its serviceability.	**Bar**	A structural element, usually oblong, used, for example, in trusses.
Angle of friction	Maximum possible angle between the contact force of two bodies and the normal axis of the contact area.	**Barrel vault**	A cylindrical vault generated by the translation of an arch along a rectilinear axis, perpendicular to its plane.
Anisotropic	Adjective to describe a structure or material that has different properties in different directions.	**Barycenter**	See *center of gravity*.
Apse	Semicircular or polygonal structure placed at the end of the nave of a church.	**Barycentric axis**	The line that joins the centers of gravity of the sections of a bar.
Arch	A load-bearing structural element with a curved shape, essentially used to carry loads by compression.	**Beam**	A structural element with a prismatic form mainly subjected to bending.
Arch stone or quoin	A block of stone with two sides on converging planes, used to construct an arch or a vault. By analogy, prefabricated segment of a linear structure.	**Beam grid**	A structure formed by crossed beams.
		Bending	The state of internal forces of compression and tension that causes the curving of a structural element.
Arch-buttress	An inclined masonry arch capable of partially countering the thrust of vaulted structures and stabilizing a construction.	**Bending moment**	The combination of tensile and compressive internal forces causing a curvature.
Arch-cable	A structure composed of two funicular members, one in tension, the other in compression, in which the thrusts on the supports compensate for each other.	**Bracing**	A structural element designed to carry transverse loads or stabilize compressed elements.
		Buckling	The major, sudden movement of a slender member under the effect of axial compression that causes instability.
Architrave	In a building, a horizontal beam supported by columns.	**Builder**	Person who has the training and expertise to complete a construction project, from the study

phase to calculation to construction. This activity is currently divided between architect and civil engineer.

Buttress A structural element in the form of a trapezoidal column used to carry and transmit inclined thrusts to the ground.

Buttressing element A structural element that opposes the displacement of the structure. For example: arch-buttress or tie in an arch-cable.

Cable A structural element stressed essentially by tension; a constructive member composed of multiple threads.

Cable beam A plane structural system composed of load-bearing cables, stiffening cables and connecting elements.

Cable beam A plane structural system composed of load-bearing cables, pretensioning cables and connecting elements.

Cable network A system of cables in space obtained by crossing two groups of cables in such a way as to form a load-bearing surface. One group of cables is load-bearing, while the other is for stiffening.

Cable-stayed Said of a structural element or structure supported by one or more stay cables.

Cantilever The part of a structure that extends beyond the outermost support.

Carrying capacity The load supported by a structure over and above its own weight and permanent loads.

Catenary curve The curve formed naturally by a cable of constant section under its own weight supported at its two ends.

Center of gravity The barycenter, point of application of the resultant of the gravitational forces acting on a body.

Centering A temporary framework to support the stones during the construction of an arch or a vault.

By analogy, temporary structure with the same function in the phase of casting the concrete.

Characteristic strength The statistically reliable strength.

Chord The upper or lower longitudinal member of a truss.

Clamp A support in which rotation is impeded.

Coefficient of friction The ratio of the force of friction between two bodies and the normal force between them.

Coefficient of thermal expansion The physical constant that defines the relationship between temperature variation and the deformation it causes.

Collapse The failure of a structure under the effect of the actions.

Column A vertical structural element able to carry compressive forces.

Compact To compress a soil to make it more stable.

Compression The state of being compressed; the internal force that causes shortening.

Compressive force An axial force that tends to shorten a member.

Compressive internal force An internal force that causes the shortening of a material or a structural member.

Compressive stress Stress that generates a shortening of the material.

Concentrated load The load acting on one point of a structure.

Conoid A doubly curved surface produced by the rotation of a generatrix curve around an axis.

Console A horizontal structural element in which one end is free and the other is clamped on the support.

Core Stiff vertical member that carries the horizontal loads acting on a building and thus stabilizes it.

Corrugated	Said of a body with one or more rippled or undulated surfaces.	**Diagonal**	An oblique element connecting the lower and upper chords of a truss.
Creep	The gradual increase in the time of deformation due to the action of stresses in concrete and wood.	**Dimensioning**	The determination of minimum size, materials and constructive details for a structural element on the basis of constructive considerations or of calculations.
Cremona diagram	A graphical construction obtained by combining multiple force polygons.	**Direction**	A line that defines the orientation of the action of a force in a plane or in space.
Crest	The summit of a pitched roof.	**Displacement**	The movement of a body.
Crest line	The peak of a folded structure, a shell or a membrane formed by multiple elements.	**Distributed load**	The load spread over a given length or area.
Critical deflected shape	The form assumed by a structure under its critical load.	**Dolomite**	Sedimentary rock formed of calcium and magnesium carbonate.
Critical load	The load under which a bar in compression develops excessive lateral deformation and loss of stability.	**Dome**	A three-dimensional structure with a doubly curved arched form, generated by the rotation of an arch around a vertical axis passing through the key.
Curing	The set of techniques used after the casting of concrete to prevent the formation of cracks.	**Ductile**	Said of a material or structural element capable of being highly stretched or deformed before failure.
Deck	The structure that provides continuous support for the circulation surface of a bridge.	**Ductility**	The capacity of a material or a structural member to undergo plastic deformation before failure.
Deck slab	The slab of a bridge on which the loads from persons and/or vehicles act.		
Deep beam	Beam with a very low slenderness ratio, of less than 3.	**Dynamics**	The branch of mechanics concerned with bodies in motion.
Deformation	A change of form, occuring as either a lengthening, a shortening or a curvature.	**Earthquake**	Rapid, sudden movement of the earth's crust that causes alternating vertical and horizontal accelerations.
Design load	The load for which dimensioning is performed, corresponding to the sum of the loads amplified by the load factors.		
Design strength	The strength used for the dimensioning, corresponding to the characteristic strength decreased by a safety factor.	**Eccentricity**	The distance between a point and an axis; by analogy, that also between the line of action of an internal force and the axis of the structural element to which it is applied.
Design value of the internal force	The internal force in a structural element amplified by the corresponding load factors.	**Edge cable**	A cable placed at the edge of a cable network or membrane to carry internal forces from load-bearing and stiffening cables.

Effective length — The length of a hypothetical column with the same critical load as a column with two hinges at its ends; the length comprised between two points of contraflexure of the deflected shape of a compressed bar.

Elastic material — A material whose deformation is reversible.

Elasticity — The property of a material or a solid body that describes its ability to return to its initial form after deformation.

Envelope curve — The curve that encloses the funicular polygons of different load configurations.

Equilibrium — A situation in which a body or a free body is at rest, and has no tendency to move.

Extrados — The external surface of an arch or a vault; by analogy, the upper surface of a structure.

Extrusion — The process of production of prismatic bars consisting by the forcing of a material through a shaped die.

Factored load — See *design load*.

Failure — A state in which, once reached, a material or a structure loses its strength.

Fan vault — A structure with a surface formed by multiple conoids (doubly curved surfaces produced by the rotation of a generatrix curve around an axis) positioned next to each other.

Fatigue — Failure caused by repetition of many periodic stresses.

Fixed support — An element that sustains a structure or structural element by constraining it along two axes and permitting its rotation.

Flexural stiffness — The stiffness of a bent beam or slab.

Force — Physical action capable of deforming a body, or modifying its movement, direction or velocity.

Force polygon — The graphic representation of a set of forces in equilibrium.

Force reduction — The replacement of multiple forces by their resultant.

Fragility — Characteristic of a material or a structural element that breaks suddenly, without plastic deformation.

Frame — Plane structural element composed of a more or less horizontal member (beam) stiffly connected to two or more vertical members (columns or piers).

Free body — A part of a system, isolated from the rest, with all the forces and internal forces acting on it.

Free span — The space between two consecutive supports.

Friction — Force that opposes the relative slippage between two bodies in contact. The frictional force depends on the nature of the contact surfaces.

Funicular arch — A curve that identifies the axis of an arch that, with a given configuration of loads, is stressed only by compression.

Funicular polygon of the loads — The shape of a structure simply in tension or in compression, in equilibrium with the loads.

Generatrix — A line generating a surface by translation or rotation around an axis.

Geometry of equilibrium — The geometry of a cable or a system of cables for which the internal forces are in equilibrium with the loads and the forces at the supports. This corresponds to the funicular polygon of the loads.

Glue-laminated wood — Material composed of layers of lumber glued on top of one another to make structural elements.

Gravitational force — The force applied to any body subjected to the gravitational field. Synonym: *weight*.

Groin vault — A vault resulting from the intersection of two or more barrel vaults in which the lower parts have been eliminated.

Hanger — A cable, generally vertical, used to suspend a load from the main structure.

Hinge	A mechanical element that holds two structural members together and permits reciprocal rotation with respect to an axis or a point.	**Load factor**	The factor by which loads are increased to consider their statistical variation for the dimensioning of structures.
Hyperbolic paraboloid	A rotational surface generated by a set of skew straight lines, whose section can be a parabola or a hyperbola depending on the position of the intersecting plane.	**Load-bearing cable**	The main element of a cable structure that carries loads to the supports.
Impact	An accidental action due to interaction between a body in motion and the structure.	**Load-bearing structure**	The skeleton of a construction that carries the loads and transmits them to the supports; any structure that carries loads.
Impost	See *springer.*	**Mechanism**	An unstable system in which a cinematic movement is produced.
Inertial force	The resistance of mass to a change in its state of motion.	**Membrane**	A structural element composed of a sheet stressed exclusively by tension.
Instability	The phenomenon of the increase of transverse movement under a compressive internal force, which can lead to failure.	**Metastable**	Said of a temporary state of equilibrium.
Intensity	The magnitude of a force or an internal force.	**Midspan**	The middle of the span.
Interaxial	The distance between the axes of similar elements.	**Mobile support**	An element that sustains a structure or structural element by constraining it along one direction and permitting its movement along the direction perpendicular to the constrained direction.
Internal force	A force acting on the section of a free body. A force acting within a structural element. Also called action effect.		
Intrados	The lower surface of an arch; by analogy, the lower surface of a structure.	**Modulus of elasticity**	The ratio between stress and specific strain of a material, which defines its stiffness. Also called Young's modulus.
Isotropic	Said of a material or structure in which the properties are the same in all the directions.	**Moment**	The product of a force times a distance.
Key	The highest part of an arch or a vault.	**Moment of inertia**	The magnitude of a section that quantifies its bending stiffness. Equal to the sum of the product of each elementary area, obtained by subdivision, by the square of its distance to an axis located in the same plane.
Lamination	The process of production of metal bars or sheets, hot or cold, through passage between two smooth or grooved cylinders.		
Lenticular	Said of objects having the form of a biconvex lens.	**Nave**	Each of the longitudinal spaces into which a church or a similar building is divided, by walls or columns.
Line load	A load uniformly distributed along a line, or along the axis of a linear element.	**Newton**	A unit of measure of a force or internal force (symbol [N]), corresponding to 1 kg·m/s²; named after Isaac Newton, 1643-1727.
Linear	Said of the relation between two directly proportional properties.		
Load	The force acting on a structure.	**Node**	The point of convergence of the bars of a truss.

Non-structural weight	The weight of all the permanent elements that are not part of the load-carrying structure.	**Poisson's ratio**	The characteristic of a material that defines the ratio between transverse and longitudinal deformation of an element strained by an axial force.
Notch	An incision made in an element.		
Ordinate	The vertical axis of a graph.	**Post**	A vertical bar of a truss that connects the upper and lower chords.
Pascal	A unit of measure of pressure (symbol [Pa]), corresponding to 1 N/m², utilized as unit of measure of a stress or a load distributed on an area; named after Blaise Pascal, 1623-1662.	**Pressure line**	Line along which the resultant of the internal force of a compressed element acts. It corresponds to the funicular polygon of the loads.
Path of structures	The sequence of this book. The path of structures progresses from the simplest to the most complex structure.	**Prestressing**	The construction system for the introduction of compressive forces in reinforced concrete by means of steel cables under tension.
Pavilion vault	A vault resulting from the intersection of two or more barrel vaults in which the upper parts have been eliminated.	**Prestressing cable**	A high-strength steel cable, generally inserted in concrete and tensioned to generate compressive internal forces beneficial for the behavior of the structure.
Pier	The generally vertical structural element that supports a bridge.	**Pretensioning cable**	A cable strained to introduce a favorable tension in the load-bearing cables in order to stabilize them. See stiffening cable.
Pillar	The vertical structural element that supports the structure of a building.		
Pinnacle	A heavy element placed at the top of a buttress.	**Purlin**	A secondary beam that transmits the loads of a roof to the main load-bearing structure.
Plastic deformation	Irreversible deformation.	**Rampant arch**	See *arch-buttress*.
Plastification	A phase of mechanical behavior of certain materials, or of a structure, in which the deformations are not completely reversible.	**Reinforcement**	See *reinforcing bar*.
		Reinforcing bar	A steel bar used in reinforced concrete to pick up tensile forces.
Plate	See *slab*.	**Relaxation**	The gradual diminishing of stresses over time in an element of concrete, wood or steel subjected to imposed deformations.
Pneumatic structure	A structure composed of a membrane (and sometimes of pretensioning cables) in which the internal overpressure of the air carries the loads.		
		Resistance factor	The factor by which the strengths of materials are reduced to consider their statistical variation for the dimensioning of structures.
Point of action	The point in which a force operates.		
Point of contraflexure	The point of a curve at which there is an inversion of the curvature.	**Resultant**	The force whose action is equal to that of a given system of forces.
Pointed arch	An arch formed by two semi-arches or by multiple bars resting on one another; the tangents at the key of the arch form a more or less acute angle.	**Retaining masonry**	Masonry placed over an arch or a vault, with a stabilizing function.
		Reticular	Having the form of a truss.

Rib	The beam-shaped reinforcement of a slab or a shell.	**Skylight**	An opening in the roof of a building that allows natural light to enter the interior.
Rise	The vertical distance between the key and the springing line of an arch, or the vertical distance between the lowest point of a cable and its springing line.	**Slab**	A flat structural element mainly loaded perpendicularly to its plane.
Roof	A structure or surface that has the function of covering and protecting a construction.	**Slenderness**	In an arch or a cable, the ratio between the span and the rise; in beams, frames and slabs, the ratio between span and height.
Round arch	An arched structure with a semicircular shape whose rise is equal to half its span.	**Sliding**	The relative movement parallel to the surface of contact between two bodies.
Section	The area obtained by intersecting a volume or a member with a plane. In the case of linear structures (cables, arches, beams), this section is generally perpendicular to the barycentric axis.	**Span**	The distance between two consecutive supports of a structural element.
		Spandrel	A masonry part placed over an arch or a vault.
Self-weight	The weight of the load-bearing structure.	**Spherical sector**	The part of a sphere intersected by two planes passing through the same axis of symmetry. The form of some domes.
Service load	The load acting on a structure at the serviceability limit state.	**Springer**	The base of an arch or a vault; the place where the curvature starts. It is also the contact zone between the support and the arch or vault.
Serviceability limit state	The state corresponding to the utilization of a structure.		
Settle	To move, as in the case of a structure resting on ground that yields.	**Springing line**	The straight line that connects the springers of an arch or a cable.
Shape	Form, overall appearance.	**Stability**	The property of a body or a structure allowing it to remain in its position of equilibrium. Branch of statics devoted to the study of stability and its conditions.
Shear force	An internal force acting in the plane of the section of a structural element it tends to shear.		
Sheet metal	A thin metal element produced by lamination.	**Static determinacy**	Property of a structure whose internal forces can be determined on the sole basis of the geometry and the acting loads.
Shell	A slender three-dimensional structure with double curvature, essentially stressed by compression.	**Statically determinate**	Said of a structure in which the internal forces can be determined on the sole basis of the geometry and the acting loads.
Shoring	The support system composed of compressed elements, generally temporary. Shoring is, for example, used to support the formwork of a concrete structure before setting of the concrete.	**Statically determinate structure**	A structure for which the internal or external constraints (supports) are equal in number to those strictly needed to guarantee stable equilibrium.
Shortening	Deformation caused by compressive force.	**Statically indeterminate**	The structural property in which the internal forces cannot be determined, knowing only the geometry of the structure and the loads acting
Shrinkage	Diminishing of length of wood or concrete, mainly due to reduction of humidity.		

on it, because the internal forces also depend on the mechanical behavior of the elements and their movement or deformation.

Statically indeterminate structure   A structure for which the internal or external constraints (supports) are in greater number than strictly needed to guarantee stable equilibrium.

Statics   The branch of mechanics regarding bodies in a state of equilibrium without movement.

Stay cable   A tie, generally in steel, used to suspend roofs or bridge decks, or to anchor forces to the ground.

Stiffened   Said of a structural element to which further structures are added to diminish its deformations.

Stiffener   A stiffening element used generally in steel constructions to prevent local buckling of slender parts.

Stiffening beam   A beam that stiffens and stabilizes an arch or a suspended structure.

Stiffening cable   A member of a structure of cables whose function is to increase the stiffeners or stabilize the structure.

Stiffness   The property of a structural element or material, expressed as the ratio between load or applied stress and the resulting deformation or unit deformation.

Strain   The ratio between the deformation and the original length. Also called the specific deformation.

Strain hardening   The phase of the mechanical behavior of steel that follows yielding, marked by a further increase of stress.

Strength   The maximum value of the force or stress a structural element or material can sustain before failure.

Stress   A quantity that indicates the intensity of the internal force in a material, defined as internal force per unit area. Symbol: σ (sigma).

Structural mechanics   The branch of physics that studies the behavior of structures.

Structural model   A simplified mechanical representation of a structure to study its behavior.

Strut   A structural element subject to compression.

Stud   Mechanical element that functions as the axis of rotation of a hinge.

Suction   A negative pressure on a surface exerted, for example, on the downwind side of a building.

Support   An element that sustains a structure, which can be of various types: fixed, sliding or clamped.

Tensile force   An axial force that tends to lengthen a member.

Tensile internal force   An internal force that causes the lengthening of a material or a structural member.

Tensile stress   Stress that generates a lengthening of the material.

Tension   The action of pulling, the force that causes lengthening.

Tent   The spatial structure of textile material strained only by tension. Similar to a cable network in which the cables touch one another.

Thermal expansion   Increase in the length, area or volume of a body due to an increase in temperature.

Thrust   The horizontal component of the force at the support of an arch or a vault.

Tie   A structural element subjected to tension.

Torque   The internal force that produces the rotation of a structural element around its longitudinal axis.

Torsion   The act of twisting, the force that causes a rotation around the longitudinal axis of a structural member.

Tower   A vertical structure of great height with respect to its base.

Truss   A structure composed of bars arranged according to a triangular grid whose members are mainly subjected to compression and tension.

Ultimate limit state   The limit state corresponding to the reaching of the ultimate (maximum) strength of a structure.

Uniformly distributed load   A distributed load with the same intensity throughout the entire length or area.

Unstressed   Adjective employed to describe a structural element without internal force and therefore not participating in the work of the load-bearing structure.

Variable load   A load with intensity that varies over time.

Vault   A structure whose surface has an arched form with a single curvature.

Vector   The segment of an oriented straight line that forms a mathematical entity; a magnitude that indicates a direction and an intensity.

Wall   A vertical planar structural element.

Wall beam   See *deep beam*.

Wind thrust   The load caused by wind.

Yielding   A phase of the behavior of ductile materials when the internal forces exceed the elastic limit and induce plastic deformations. We then speak of *yield strength*.

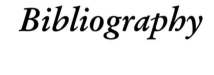

Bibliography

Gordon J. E., *Structures, or why things don't fall down*, Plenum Press, New York, 1978

Gordon J. E., *The Science of Structures and Materials*, Scientific American Books, New York, 1988

Nervi P. L., *Structures*, F.W. Dodge Corp., 1956, 118 pages

Pizzetti G., *Principi statici e forme strutturali*, UTET, Torino, 1980

Salvadori M., Heller R., *Structures in Architectures*, Prentice-Hall, Englewood Cliffs, USA, 1963

Torroja Miret E., *Philosophy of Structures*, University of California Press, 1958

Qualitative and intuitive approach to the understanding of the behavior of structures

Allen E., Zalewski W., *Form and Forces: Designing Efficient, Expressive Structures*, John Wiley and Sons, 2010

Zalewski W., Allen E., *Shaping Structures: Statics*, Ed. John Wiley and Sons, New York, 1998

Approach based on graphical statics

Ackermann K., *Tragwerke in der konstruktiven Architektur*, Deutsche Verlags-Anstalt, Stuttgart, 1988

Billington D. P., *The Tower and the Bridge, the new art of structural engineering*, Princeton University Press, Princeton, New Jersey, 1983

Charleson A., *Structure as Architecture: a Source Book for Architects and Structural Engineers*, Architectural Press, 2005

Deswarte S., Lemoine B., *L'architecture et les Ingénieurs*, Le Moniteur, 1997

Margolius I., *Architects + Engineers = Structures*, Wiley-Academy, 2002

Mark R., *Light, Wind and Structure*, MIT Press, 1990

Mock E. B., *The Architecture of Bridges*, The Museum of Modern Art, New York, 1949

Picon A., *L'art de l'ingénieur, constructeur, entrepreneur, inventeur*, Editions du Centre Pompidou/Le Moniteur, Paris, 1997

Robbin T., *Engineering a New Architecture*, Yale University Press, New Hawen and London, 1996

Siegel C., *Structure and Form in Modern Architecture*, R. E. Krieger Pub. Co., 1975

Structures viewed from a perspective between architecture and engineering

Angerer F., *Surface Structures in Building: Structure and Form*, Reinhold Pub. Corp., 1961

Frei O., *Zugbeanspruchte Konstruktionen*, Ullstein Verlag, 1966

Heinle E., Leonhardt F., *Towers: a Historical Survey*, Rizzoli, 1988

Heinle E., Schlaich J., *Kuppeln aller Zeiten, aller Kulturen*, Deutsche Verlags-Anstalt, Stuttgart, 1996

Ishii K., *Membrane Structures in Japan*, SPS Publishing Company, Tokyo, 1995

Joedicke J., *Shell Architecture*, Reinhold, 1963

Leedy W. C., *Fan Vaulting: a Study of Form, Technology, and Meaning*, Arts+Architecture Press, 1980

Types of structures

Leonhardt F., *Brücken/Bridges, Aestetik und Gestaltung, Aesthetics and Design*, Deutsche Verlags-Anstalt, Stuttgart, 1982

Mierop C., *Skyscrapers, higher and higher*, Norma, Paris, 1995

Rice P., Dutton H., *Structural Glass*, Spon Press, 1995

Van Beek, G. W., *Arches and Vaults in the Ancient Near East*, Scientific American, 257, 1987, no. 1: 78-85

Thoughts and perspectives of engineers and architects

Freyssinet E., *Un amour sans limite*, Editions du Linteau, Paris, 1993

Nervi P. L., *Scienza o arte del costruire?*, 1945, (édition récente: Cittàstudi Edizioni, Milano, 1997)

Nervi P. L., *Nuove strutture*, Edizioni di Comunità, Milano, 1963

Rice P., *An Engineer imagines*, Artemis, London, 1994.

Wachsmann K., *The Turning Point of Building: Structure and Design*, Reinhold Pub. Corp., 1961

Engineers and architects

Anderson S., *Eladio Dieste: Innovation in Structural Art*, Princeton Architectural Press, 2004

Bill M., *Robert Maillart*, Artemis Verlag, Zürich, 1949

Billington D. P., *Robert Maillart's Bridges, the art of engineering*, Princeton University Press, New Jersey, 1979

Brownlee D.B., De Long D.G., *Louis I. Kahn: in the realm of architecture*, Museum of contemporary Art, Los Angeles, 1991

Buchanan P., *Renzo Piano Building Workshop*, Phaidon, London, 1993-1998

Candela F., Segui Buenaventura M., *Felix Candela, arquitecto*, catalogue d'exposition, Instituto Juan de Herrera, Madrid, 1994

Chilton J., *Heinz Isler, The Engineer's Contribution to Contemporary Architecture*, Thomas Telford Publishing, 2000

Desideri P., *Pier Luigi Nervi*, Zanichelli, Bologna, 1979

Dieste E., *Eladio Dieste 1943-1996*, Junta de Andalucia, Sevilla, 1998

Dini M., *Renzo Piano*, Electa, Milano, 1983

Dunster D., *Arups on Engineering*, Ernst&Sohn, Berlin, 1996

Fernandez Ordonez S. A., *Eugène Freyssinet*, 2C Editions, Barcelona, 1979

Fernandez Ordonez S. A., Navarro Vera J. R., *Eduardo Torroja, Ingeniero*, Ediciones Pronaos, Madrid, 1999

Guidot R., Guiheux A., *Jean Prouvé, constructeur*, Editions du Centre Pompidou, Paris, 1990

Holgate A., *The art of structural engineering the work of Jörg Schlaich and his team*, Axel Menges, Stuttgart, 1997

Imbesi G., Morandi M., Moschini F., *Riccardo Morandi, Innovazione, Tecnologia, Progetto*, Gangemi Editore, Roma 1995

Killer J., *Die Bauwerke der Baumeister Grubenmann*, Birkhäuser Verlag, Basel, 1985

Loyrette H., *Gustave Eiffel*, Deutsche Verlagsanstalt, Fribourg 1985

Marrey B., *Nicolas Esquillan, un ingénieur d'entreprise*, Picard éditeur, Paris, 1992

Morandi R., Boaga G., Boni B., The Concrete Architecture of Riccardo Moranti, Praeger, 1966

Moreyra Garlock M. E., Billington D. P., *Félix Candela: Engineer, Builder, Structural Artist*, Yale University Press, 2008, 207 pages

Nardi G., *Angelo Mangiarotti*, Maggioli Editore, Rimini, 1997

Nerdinger W., *Frei Otto: Complete Works : Lightweight Construction, Natural Design*, Birkhäuser, 2005

Pawley M., *Norman Foster: a Global Architecture*, Universe, 1999

Pedretti C., *Leonardo, Architect*, Rizzoli, 1985

Pfeiffer B.B., *Frank Lloyd Wright*, Taschen, Köln, 1991

Polano S., *Santiago Calatrava*, Electa, Milano, 1996

Rui-Wamba J. *et al.*, *Eugène Freyssinet, Un ingeniero revolucionario*, Fundacion Esteyco, 2003

Schulze F., *Mies van der Rohe*, Museum of Modern Art, New York, 1989

Stüssi F., *Othmar H. Amman, Sein Beitrag zur Entwicklung des Grossbrückenbaus*, Birkhäuser Verlag, Basel, 1974

Suchov V., Graefe R., Gappoev M., Pertschi O., *Die Kunst der sparsamen Konstruktion*, Deutsche Verlags-Anstalt, Stuttgart, 1992

Villa A., *Silvano Zorzi*, Electa, Milano, 1995

Von der Mühl H.R., *Kenzo Tange*, Zanichelli, Bologna, 1981

History of construction

Adam J.-P., Mathews A., *Roman Building: Materials and Techniques*, Routledge, 2003

Jesberg P., *Die Geschichte der Ingenieurbaukunst*, Deutsche Verlags-Anstalt, Stuttgart, 1996

Lamprecht H.-O., *Opus caementicium, Bautechnik der Römer*, Beton-Verlag, Düsseldorf, 1984

Peters, T. F., *Transition in engineering, Guillaume Henri Dufour and early 19th century cable suspension bridges*, Birkhauser, Basel, 1987

Pfammatter U., *Building the future: Building Technology and Cultural History from the Industrial Revolution until Today*, Prestel, 2008

Structures discussed in this book

The Golden Gate Bridge, Highway and transportation district, San Francisco, 1937/1987

Bellinelli L., *Louis I. Kahn: The Construction of the Kimbell Art Museum*, Skira, 1999

Boesiger W., *Le Corbusier Œuvre complète, 1952-1957*, Les Editions d'Architecture, Zurich, 1970

De Bures C., *La tour de 300 mètres*, Editions André Delcourt, Paris, 1998

Eggen A. P., Sandaker B. N., *Steel, Structure, and Architecture*, Whitney Library of Design, 1995

Le Corbusier, *My Work*, Architectural Press, 1960

Lyall S., *Masters of Structures*, Kohlhammer, Stuttgart, 2002

McKean J., *Crystal Palace, Paxton and Fox*, Phaidon, London, 1994

Piano R., Rogers R., *Du Plateau Beaubourg au Centre Georges Pompidou*, Le Moniteur, Paris, 1987

Saarinen E., *Il terminal dell'aeroporto internazionale Dulles*, Jaca Book, Milano, 1994

Thompson, D'Arcy W., *On Growth and Form*, The complete revised edition, Dover Publications, 1992

Vischer J., Hilbersheimer L., *Beton als Gestalter*, Julius Hofmann Stuttgart, 1928

Photography credits

The sources for the photographs reproduced in this work are summarized below, the numbers indicating the page and the order of the appearance on the page. When a name is followed by a date, the complete reference for the image can be found in the bibliography. In other cases, the copyright holder is simply mentioned.

Every time we were able to identify a copyright owner, a request to reproduce the image was sent to that person. However, several requests remain unanswered at the time of the publication of the first edition of this book. Also, we were simply not able to identify a certain number of rights holders: for those who identify images that should have been acknowledged, please contact the Presses polytechniques et universitaires romandes, who will take care to cite the source in a future edition. The author and the editor thank all those who have agreed to give their authorization for reproduction in this work, and thank also those whom we were unable to identify in this volume for their comprehension.

Acier-stahl-steel, 1974 – 103/4a-c

Ackermann, 1988 – 157/3

D'Arcy, 1992 – 177/1a-d

Bill, 1949 – 75/3, 79/2b-c, 82/1a-b

Boaga, 1984 – 59/4

Boesiger, 1970 – 54/2a-d, 90/5, 102/2a

Brownlee, 1991 – 205/1

E. Brühwiler, – 4/2

Buchanan, 1993 – 103/2a-b (Photos Shungi Ishida and Gianni Benengo Gardin)

Candela, 1994 – 107/2a-c, 108/2a-b, 108/3

Construire en Acier, Centrale Suisse de la Construction Métallique, Zurich, 2002 – 97/3a-b

Conzett J., 198/3

L. Curran, Maecenas, http://wings.buffalo.edu/AandL/Maecenas – 190/1

De Bures, 1998 – 153/1a-b, 187/1a+c (Fonds Eiffel, ENPC, E. Monod)

Desideri, 1979 – 92/2, 185/2a-b

Deswarte, 1997 – 80/3, 183/1b

Dieste, 1998 – 90/2b

Dunster, 1996 – 75/2a

Eggen, 1996 – 89/1b (photo Arne Peter Eggen)

B. Elliott & H. Monroe – 57/3a-d

R. Favre, PPUR – 93/2

R. Feiner – 193/1 (photo Ralph Feiner)

Fernandez Ordonez, 1979 – 114/1, 181/1, 181/1a-b

A. Flammer – 200/3

Frei, 1966 – 55/3c

Golden Gate, 1987 – 47/3a-b

S. Guandalini – 59/2a-b, 206/1, 209/1

Guidot, 1998 – 158/4a, 189/1a-b (© Galerie Patrick Seguin)

Heinle et al., 1996 – 95/1a-b, 96/3a-c, 98/2, 100/1a-b

Heinle et al., 1990 – 175/4

Holgate, 1997 – 57/1a-b, 67/1, 104/4a-b, 104/5a-b, 110/2a-c

Imbesi et al., 1995 – 178/3

INCO Spa, Milan – 186/1

Ishii, 1995 – 68/3

H. Isler – 105/2, 106/1, 106/2

Jesberg, 1996 – 73/1a-b (Hart, Poleni)

Kahn, 1997 – 109/2

C. Kerez – 199/2

J.-F. Klein – 219/2

Le Corbusier, 1960 – 102/2b

Leedy, 1983 – 92/3a-b

Leonhardt, 1982 – 180/2

Loyrette, 1985 – 129/2, 150/3a

Lyall, 2002 – 228/1 (photo Grant Smith)

Mark, 1990 – 92/1a-b, 113 (©MIT Press)

McKean, 1994 – 141/1a-b

Mierop, 1995 – 119/4, 150/1b

A. Muttoni – 4/1, 65/1, 81/2, 83/1, 84/3, 85/3, 85/4, 91/3, 101/2, 114/4a, 119/1, 119/2, 139/3b, 145/6, 150/1a, 150/2c, 193/3, 207/1, 222/2, 223/4, 228/2

Nervi, 1965 – 48/2a-b, 90/2a, 97/1a-c, 153/1a-b, 184/1b, 205/2, 228/3

Nervi, 1963 – 177/2, 177, 192/1a-b, 200

Palladio, 1570 – 140/1

P. Pasinelli – 22/2a-d, 165/1, 218/1a-b, 218/2a-b, 227/2a-b

Pawley, 1999 – 219/3

Foto Pedroli, Chiasso – 117/2

Pfeiffer, 1991 – 208/2 (© 2004, ProLitteris, Zurich)

Pfeifer Seil- und Hebetechnik GmbH, D-87700
Memmingen – 58/1, 65/2

Picon, 1997 – 47/2b (photo Erza Stoller), 120/4, 64/1
(coll. Mnam-Cci, Ph. Baranger), 66/1a-b (photo
Horst Berger), 68/1 (photo Horst Berger), 68/4a-c
(Taiyo Kogyo Corp.), 72/2 (Severud Assoc.), 79/2a
(Fonds Eiffel), 80/2 (Bibl. EPFZ), 93/3a-c (coll.
EPAD, photo Jean Biaugeaud), 105/1 (coll. David
Billington), 118/2 (AKG Photo), 145/2 (Deutsches
Museum München), 152/3 (Deutsches Museum
München), 158/2a-b (Fuller and Sadao PC), 188/4
(Fonds Prouvé)

Pizzetti, 1980 – 73/2a-b

S. Ranshaw - www.SailPhoto.co.uk &
www.Ranshaw.co.uk – 215/1

Rice, 1995 – 56/1, 56/2

Robbin, 1996 – 65/3, 67/4, 68/2

Saarinen, 1994 – 47/2a

Schulze, 1989 – 140/2, 200/2a-b

J. Schwartz – 199/1

Studio Vacchini – 193/2

Suchov, 1990 – 75/1a-c, 102/3a (photos R. Graefe)

P. Thürlimann – 150/2b

Van Beek, 1987 – 89/3, 90/4

Villa, 1995 – 180/1a-b, 186/2,3

Vischer, 1928 – 90/3, 208/1

von der Mühl, 1981 – 58/3

Wachsmann, 1962 – 158/2a-b, 158/3a-b, 158/4b

W. & O. Wright, Library of Congress, Washington DC
– 141/4

J.-L. Zanella – 217/1a-c

Index

A

action, 12
analysis
– (complete, of a truss), 133
– of diagonals, 137
– (general, of trusses), 129, 135
arches, 71
– auxiliary, 149
– catenary, 72
– crossed, 97
– funicular, 72
– full centered, 87
– (gridshell structures of), 110
– (influence of variable loads on), 73
– parabolic, 72
– (stabilization of), 74
– statically determinate, 78
– statically indeterminate, 78
– (the instability of), 74
– three-hinged, 80
– two hinged, 82
– with one hinge, 83
– with stiffening beam, 75
– with ties, 114
– without hinges, 83
arches and cables, similarity between 73
arch-cables, 111

B

bars,
– (compressive strength of), 220
– most strained among chords, 135
– stiffening, 75
beam(s), 161
– bi-clampled, 180
– continuous beams, 179
– (curvature of a), 165
– cable beams, 54
– Gerber, 178
– (grids of), 204
– in reinforced concrete, 164
– simple, 174
– (simple bending of), 165
– (strength of), 165
– (stiffness of), 168
– stiffening, 57, 75 (see *cables with* or *arches with*)

– (under concentrated loads), 174
– (under distributed loads), 174
– Vierendeel, 192
– with cantilevers, 175
beam bending, 165
beam grids, 204
behavior
– elastic, 17
– fragility, 22
– linear, 17
– of concrete, 23
– of glass, 22
– of steel, 20
– of stone, 23
– of wood, 24
– plastic, 19
bending moment, 134, 165, 234
– simple, 165
– simple of a beam, 165
buckling, 227

C

cable(s), 35
– as stabilizers, 56
– auxiliary, 44
– beams, 54
– (deflections of), 51, 52
– networks, 64
– (influence of load position), 40
– stiffening cables, 54, 65
– (subjected to uniform loads), 45
– (support of), 49
– suspension, 47, 65
– (systems of cables in space), 63
– with flexural stiffness, 58
– with multiple non-vertical loads, 43
– with struts, 116
– with stiffening beam, 57
– with two vertical loads, 41
cable-stayed systems, 59
catenary, 47, 72
– (arch in form of), 72
cantilevers, 175
– (beams with), 177
choice of slab thickness, 206
chords, 132
– (bars under largest stress), 135
– lower, 132

– upper, 132
comparison of materials, 24
compression, 14, 73
– (compressive stress), 14
– (effect on materials), 14
– (structures under), 71
concrete
– (behavior of), 23
– (compression resistance of), 23
– (modulus of elasticity of), 23
– (tension resistance of), 23
conditions of equilibrium, 11
– (of more than two forces), 29
conical domes, 100
Cremora diagram, 31
criterion, 26, 27
– of the serviceability limit state, 26
– of the ultimate limit state, 27
critical load, 218
cylinders, 109

D
deep walls, 197
deflections, 51, 52
– allowable, 207
– of cables, 51
– caused by permanent loads, 51
– caused by variable loads, 51, 53
– caused by temperature variation, 52
diagonals, 132, 137
– (analysis of), 137
– (internal forces in), 138
– K diagonals, 142
– N diagonals, 140
– V diagonals, 139
– (under compression), 138
– X diagonals, 140
dimensioning, 26, 28
– of cables, 49
– of slabs, 206
– (internal force in), 27
– (resistance factors in), 27
domes, 95, 100, 105
– conical shaped, 100
– geodesic, 104
– hyperboloid of revolution, 101
– metallic, 97
ductility, 22

E
effective depth, 165
effective length, 221
elastic limit, 20
elongation, 16
equilibrium, 11
– conditions of, 11
–(of more than two forces), 29

F
fatigue, 29
flanges, 171
flat slabs, 212
forces, 7, 29, 32
– of inertia, 9
– of gravity, 9
– (polygon of), 30
– transmission of, 13
form or shape, 82, 99
– ideal for two-hinged arches, 82
– ideal for three-hinged arches, 79
– of domes, 99
– of trusses, 145
fragility, 22
frames, 183
– stacked, 191
– multistory side-by-side, 192
– three hinged, 186
– two hinged, 185
free-body diagram, 11
friction,
– (angle of), 31

G
geometry of cables (influence of), 39
Gerber trusses, 152, 178
glass, 22
– (behavior of), 22
gravity, 9
grid domes, 104
gridshell stuctures, 110

H
hinge(s),
– two-hinged arches, 82
– three-hinged arches, 80
– two-hinged frames, 185
Howe system, 140
hyperbolic, 64

– (paraboloid), 64, 107
hyperboloids of revolution, 101
– (domes in the form of), 103

I
influence, 39, 40
– of beam structure on stiffness, 168, 172
– of dimension on internal forces in trusses, 130
– of load position on cables, 40, 49
– of support type on slab behavior, 208
– of the form on structure stiffness, 148
– of the geometry, 39
– of the load, 39
– of the position of the load on the cables, 40
– of the variable load on the arches, 73
internal force(s), 13, 15, 32, 76, 132
– axial, 13, 14, 234
– in diagonals, 132
– in dimensioning, 27
– line of action of, 76
– normal, 12
– of beams, 163
– of compression, 14
– of frames, 187
– of tension, 13, 15
– shear, 137, 234
instability of arches, 74

J
Jawerth system, 55

L
lattice of trusses, 157
law of gravitation, 9
limiting cable deflections, 53
line
– of action, 10
– of action of internal forces, 78
– of action of a force, 11
– of force, 12
linear behavior, 17
load, 9, 39, 40, 45
– concentrated, 174
– critical load, 220
– distributed (see *beams under*), 174
– (factors of), 27
– (influence of), 39
– permanent, 9, 51

– uniformly distributed, 45
– variable, 9, 51
load factors, 27
Long system, 140

M
material,
– (comparison of), 24
material quantity
– of cables, 50
– of beams, 170
– of trusses, 146
membranes, 65, 67
– pneumatic, 67
Mero system, 158
module of elasticity, 20
– of concrete, 23
– of steel, 20
moment, 232
– bending moment, 134, 234
monkey saddle surfaces, 108

N
network of cables, 64

P
parabolas, 47, 231
paraboloid
– hyperbolic, 64, 107
path of structures, 2
phase
– elastic, 19
– plastic, 19
point of application, 10
polygon,
– funicular polygon, 45
– of forces, 30
purpose of structure, 4

R
rampant arch, 33, 71, 92, 99, 113
reaction, 12
reinforcement masonry, 85
resistance factors, 27
ribbed slabs, 201
rise, 37

S

serviceability limit state (SLS), criterion of, 26
shed, 157
shells, 105
– cylindrical, 109
– downward double curvature, 106
– hyperbolic, 107
shortening, 16
slabs, 201, 205
– mushroom, 212
– ribbed, 203
– (sizing of), 205
slab thickness, 206
slenderness (l/f) ratio, 40, 49
span, 37
– equivalent span, 208
stabilizing arches, 74
stability of compressed members, 215
steel, 20, 21
– (ductile behavior of), 22
– (elasticity module of), 20
– (types of), 22
stiffness, 17, 24
– of a structure, 18
– of material, 18
stone,
– behavior of, 23
strain hardening, 20
strength, 24
– (factors of), 27
– of beams, 165
– of concrete in compression, 23
– of concrete in tension, 23
– design strength, 28
– of materials, 20
– of a rod under compression, 218
structural diagrams, 37
structural efficiency, 146, 172
structure(s),
– folded plate, 200
– load carrying, 3
– pneumatic, 67
– under compression, 71
– under tension, 38
supports, 38, 49
– of cables, 49
– fixed, 114
– sliding, 114

suspension bridges, 47
systems,
– Howe, 140
– Jawerth, 55
– Long, 140
– Mero, 158
– of cables in space, 63
– Polonceau, 145
– statically indeterminate, 128
– stable, 128
– unstable, 128
– Warren, 139
– with combined cables, 59

T

tents, 65
towers and cantilevers, 149
tension, 16
– (internal forces due to), 15
– tensile stress, 16
– (structures under), 37
thrust,
– (carrying horizontal component of), 113
– of the wind, 113, 188
transmission of force, 13
trusses, 123, 129, 157
– (analysis of), 125, 129
– (influence of depth and span on), 132
– Gerber, 152
– lenticular, 145
– lattice trusses, 157
– Long, 140
– Polonceau, 145
– space trusses, 158
– Warren, 139
– with cantilevers, 151
triple-hinged
– arches, 79
– frames, 186
types of steel, 21

U

ultimate limit state (ULS), criterion of, 27

V

variation
– of load configuration, 52
– of temperature, 52 (see *deflections due to*)

vaults, 87
– barrel, 89
– groin, 91
– fan, 92
– pavilion, 94
vector force, 10
Vierendeel beams, 192

W
wall beams, 197
Warren system, 139
wide-flange sections, 169
wood, 24
– (behavior of), 24

Y
yield strength, 20

i-structures

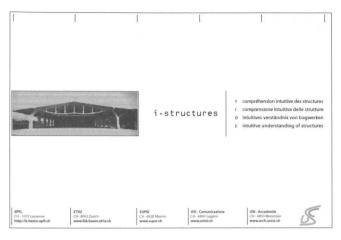

Homepage i-structures.epfl.ch

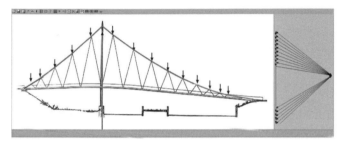

Graphical resolution of the Rosenstein I footbridge, Stuttgart, Germany, 1976, eng. J. Schlaich. (Image adopted from Fussgänger-brücken 1977-1992 von Jörg Schlaich und Rudolf Bergermann, Katalog zur Ausstellung an der ETH Zürich, Zürich, 1992)

Graphical resolution of the Lowry Millenium Bridge, Salford, UK, 2000, eng. W. Middleton. Note the force in the tension chord that is given automatically.

Our experience of the past years has shown it possible to provide a direct application of the approach outlined in this book, dedicated to the education of architects and civil engineers. By offering a rigorous application of graphical statics, students are able to thoroughly assimilate the basic principles and understand the modes of load-carrying mechanisms of structures.

However, the systematic construction of force and funicular polygons is inherently repetitive and requires an investment of time, in particular for the material of the later chapters.

As part of the Swiss Virtual Campus initiative (www.virtualcampus.ch), the i-structures online course has been developed based on the subject matter of the present book. The material is organized in a similar manner, but with a finer subdivision of the contents, to allow for more progressive and autonomous study. Only the fundamental aspects are treated, and users of the web site are referred to the book for detailed developments and specific topics.

Access to the i-structures on-line course is free for everyone after a simple registration procedure; the full content of the web site can then be consulted. Address: i-structures.epfl.ch.

The flexibility of a computer-based approach allows the subject under consideration to be dynamically changed to reflect the progression or the activity of the user. In particular, the graphical Java applet i-Cremona automatically provides the funicular polygon and the polygon of forces in a combined display. The use of background images helps the user understand the behaviour of actual structures, even complex ones.

Based on the definition of the supports and the acting forces, the applet automatically calculates the corresponding funicular polygon. Because an infinity of such polygons is possible, the user can control the shape of the polygon by specifying the initial angle at one of the supports. The results are shown in graphical form, with a numerical output when a segment of the funicular polygon is selected by the pointer.

This tool can be used to understand the behavior of existing structures and to investigate the conceptual design of new projects, as the background image can be freely uploaded by the user.

i-structures: an on-line course with tools that follow the approach of the book

Access to the on-line course

i-Cremona applet